AF615384

Topics in Statistical Mechanics and Biophysics:
A Memorial to Julius L. Jackson

(Wayne State University–1975)

AIP Conference Proceedings
Series Editor: Hugh C. Wolfe
No. 27

Topics in Statistical Mechanics and Biophysics: A Memorial to Julius L. Jackson

(Wayne State University–1975)

Editor
R.A. Piccirelli

American Institute of Physics
New York 1976

L.C. Catalog Card No. 75-36309
ISBN 0-88318-126-6

American Institute of Physics
335 East 45th Street
New York, N.Y. 10017

Printed in the United States of America

This volume is dedicated to the memory of Julius L. Jackson.

It consists of original contributions by his friends and colleagues in the major areas of his interest.

PREFACE

This collection of essays in two major areas of modern science is a memorial to the late Julius L. Jackson. Each of the authors has aimed to achieve this goal by making his essay a significant and lasting contribution to his chosen problem area. We fully expect that graduate students as well as senior scientists will find in these essays the kind of intellectual stimulation and challenge that Julius himself tried to give others while he lived.

This volume was conceived by a small committee at Wayne State University where Professor Jackson was Chairman of the Department of Chemical Engineering and Materials Sciences, and it was funded by their efforts. The group includes: Mrs. Raya Jackson; Ralph Kummler, present Chairman of Chemical Engineering and Materials Sciences; Melvin Shaw from Electrical Engineering; Alvin Saperstein from Physics; and Bob Piccirelli from Mechanical Engineering. In this undertaking we have intended to act according to the spirit of all of Julius' friends and colleagues.

The contributions fall in the areas of Julius' major scientific concern: statistical mechanics, and lately, biophysics. That these two are not exhaustive of his work is apparent from the appended bibliography. Because the authors are themselves as eclectic as Julius, we felt it safer to arrange their essays alphabetically by author rather than to group them by subject. The unselfish work of the authors in preparing their research for this volume is acknowledged with pleasure and gratitude. All their contributions were received by October 1, 1975.

Biographical material about Julius is contained both in the articles and in the following prologue. Some of the authors referred to connections with Julius' work in their background remarks and the short appended note by Fred Gornick also indicates Julius' synergistic scientific style. The prologue is an amalgam of information and anecdotes provided not only by the authors but also by a very large number of Julius' friends and colleagues in science. Thus, we acknowledge with thanks the response of all those who provided the insights into Julius' life which we have tried to reflect in the prologue.

TOPICS IN STATISTICAL MECHANICS AND BIOPHYSICS:
A Memorial to Julius L. Jackson

Table of Contents

Julius L. Jackson

1924-1974

PROLOGUE

With the sudden death of Julius L. Jackson while visiting the Weizmann Institute in Rehovot, Israel, on July 5, 1974, the physics community lost a productive, stimulating and wonderful colleague.

Professor Jackson was born of Jewish immigrant parents on 9 November, 1924, in New York City. He received degrees at Brooklyn College, Princeton, and New York University, where he earned his Ph.D. in 1950. He served as a visiting professor at the State University of Iowa prior to joining the Applied Physics Laboratory of the Johns Hopkins University as a research physicist in 1951. He also served for two years at the Office of Naval Research and in 1956 he became a research physicist at the National Bureau of Standards where he worked in the Free Radicals Program and in the Statistical Physics Section. During this period he was Fulbright Research Professor of Physics at the University of Leiden. In 1965 he joined Howard University as Professor of Physics. In 1969 he came to Wayne State University as Professor and Chairman of the Department of Chemical Engineering and Material Sciences, a post he resigned in June, 1974, in order to devote more time to teaching and research. From about 1960 on, he was consultant for several companies and was a frequent visitor to the Weizmann Institute and other research facilities. He was a Fellow of the Washington Academy of Sciences and of the American Physical Society.

Julius' research contributions are very diverse in character. He published over 50 papers in statistical physics, transport phenomena, biophysics, polymer theory and stability in dissipative systems. Among his earliest works were pioneering efforts concerning the derivation, for general quantum systems, of the relation between spontaneous equilibrium fluctuations and transport coefficients. These contributions began at APL with research on optimum linear filtering, which was of much practical interest for the design of missile guidance systems. Julius' unusual insight and ability to generalize led him to observe that the results which he and M. Yovits had found in optimum linear filtering were in fact applicable in a totally different field, namely quantum statistics. This interest in the theory of transport and the origins of irreversibility continued throughout his career. Later, he studied the basis of the autocorrelation function expressions for transport coefficients and techniques for calculating the transport coefficients of composite materials.

Another significant part of his contribution occurred during his participation in the free radical program at the National Bureau of Standards from early 1957 to the middle of 1959. There he worked on the distribution of trapped radicals, their reactions, stability, mechanisms of trapping and interaction with magnetic fields. He collaborated with many theoreticians, some of whom are authors in this volume. Perhaps even more important than his theoretical studies was the effect on the experimentalists of his probing questions and enthusiasm for the project.

Concurrently, his interest in polymers was stimulated and he and his younger colleagues and students made several contributions to the theory of diffusion in polyelectrolytic solutions. This work spanned the period both at NBS and at Howard University. There was also work on the theory of potentials in electrolytes, polymer crystallization and rubber elasticity.

His efforts in plasma physics span the same period and include several papers using the methods of statistical mechanics to calculate the distribution of fields and the thermodynamic properties of equilibrium plasmas. He also examined some aspect of the consequences of the Vlasov equation.

Of all the transport phenomena, he was most curious about particle diffusion. His physical insight in this area was keen and he used it in solving problems concerning the diffusion of electrons on the Fermi surface and of ions in polyelectrolytes. More recently he was using his understanding of transport properties in biophysical problems such as modeling chemotactic movement in bacteria and for explaining the basis of dissipative structures in biosystems.

In addition to these main threads of his interest, he also contributed to other areas of applied physics such as noise in avalanche devices, short laser pulses, and wave propagation at low pressures.

Julius enjoyed brief visits to many research centers and universities. He would immediately become involved with some problem of the group and usually succeeded in contributing to its solution. For example, in the summer of 1969 he became caught up in the efforts of the group at Livermore who were trying to further develop the theory of mode-locked lasers. The resulting paper with J. Creighton received about 60 requests for reprints from all over the world.

Julius had a unique personal and scientific style. His profound dedication to intellect was partially masked by his down-to-earth manner and his love for and sense of the absurd. Behind that enthusiasm, that wry appreciation of human achievement and failure in the attempt, was, however, a deep commitment to learning. Perhaps the most quickly perceived of his personal characteristics were his enthusiasm and generosity with others. He was delighted to discuss the work of others regardless of whether they were graduate students or senior scientists, whether they were in the next office or the next continent. His telephone was never far from his ear. He plainly showed his pleasure in describing his own work (most of which was done during the post-midnight hours) and in so doing was able to remind one that science was exciting and satisfying.

Those of us who were associated with him as colleagues in research and teaching will attest to his really exceptional physical intuition. He had a highly developed ability to translate his insight into mathematical forms which, while containing the essential elements of the phenomena, remained unencumbered with unnecessary details. This was particularly true of his feeling for random phenomena and he enjoyed finding simple but very cogent physical arguments to explain his results in such systems.

This was also illustrated during his last year at Wayne, when he worked with M. Shaw on bias induced switching transitions in crystalline and amorphous solids. According to his style, he indeed formulated an elegantly simple and tractable model which both agreed with experiment and led to new insights. After an intensive analysis of a thermally induced, negative differential conductance operating point, he spent many fretful hours seeking a physical explanation for the stability. At first no explanation satisfied his standards. This search finally led to the discovery of a missing factor of two after which a useful physical explanation was indeed possible.

Because his scientific treatment of a problem never strayed far from physical intuition, his students usually had a clear idea of the essential points in a problem, could better evaluate the result, and apply it to new situations. Many students and beginning scientists and engineers learned from him what research is all about and how to do it, and their interactions with him were profoundly useful to them. His students and colleagues will also recall with a sense of loss the clarity and force of his lectures and the stimulation of his return from one of his many trips filled with new ideas and results.

Julius Jackson served the scientific community through his research and involvement with students and professional colleagues in universities and private industry and through active membership in professional societies. Those of us who were with him at the National Bureau of Standards remember his attention and concern for the proper role of basic science there. Those at Howard University remember how deeply concerned and involved he was with the aims and problems of that University, and the contributions he made toward their solution. In his last years at Wayne he played a central and key role in the growth and maintenance of a substantial applied science community in the College of Engineering and was a source of guidance for many of us.

Julius was a warm, open human being who lived a full and satisfying life. He was very knowledgeable in music and the arts and enjoyed them enormously. While he was not a joiner, the problems of society affected him very strongly and he was a vigorous supporter of Jewish causes and Israel. He had great love and pride in his family and worked diligently to guide and nurture them. His home was a meeting ground for people of diverse talents, interests and origins. He was always ready to help others. Even as a youngster he helped Dr. S. Ajl, a newcomer from Poland in the 30's who is now a distinguished researcher. The latter remembers that he went through high school with Julius' indispensable help as an interpreter between his Yiddish and the English of his high school teachers.

Julius' many interests, dynamic personality and strength of character inspired all around him.

THE CONTRACTION OF MUSCLE AS A STOCHASTIC PROCESS

Andrew G. De Rocco*
DFCP, USAF Academy, Colorado 80840

ABSTRACT

A model is presented for the contraction of vertebrate striated muscle which is predicated largely on structural grounds and ascribes to the binding of myosin and actin subunits the primary element of force generation. The dynamics of the model is realized as a series of "states" and transitions among them which can be characterized as a biased random walk in one dimension. Two modes are described for the motion of the myosin subunits (HMM-1): (1) "inchworm," where the relative configuration of the two elements of the duplex head remains unchanged, and (2) "flip-flop," where the heads are free to undergo reversal of ordinal position. The master equation is solved in some detail for the "inchworm" mode and the solutions displayed. The mean first passage time (MFPT) is computed for the "inchworm" mode and related to the velocity of muscle contraction. For the case where both modes operate the MFPT is computed by a perturbation procedure and the effects of the "flip-flop" mode are shown to decrease the MFPT. When load dependence is included in the characterization of the elementary rates (attachment/detachment), it is shown that the stochastic theory agrees, for small loads, with the phenomenological equation of A. V. Hill relating velocity and load. A brief discussion is made of the length-tension isotherm and estimates made for the force generated per crossbridge and for the energy release necessary for unit displacement by a myosin head along the actin polymer (force $\sim 10^{-7}$ dynes; energy $\sim$ 3 kcal mole^{-1}).

INTRODUCTION

Since the original formulation of the sliding filament theory for muscle contraction [1] several mechanisms have been proposed to account for the generation of the active force of contraction. Although these theories vary considerably in their explicit physical details, they generally can be classified into two groups: those in which the force is considered to arise directly, or indirectly, from the actin-myosin interaction, [2] and those in which such an interaction, if it occurs at all, is only coincidental to a force arising from some collective property of the system. [3]

Although structural and biochemical evidence weigh heavily in favor of the first type of mechanism (the best known theoretical model is due to H. E. Huxley [2]), some evidence for the second type does exist [4] and there is some indication that an original tenent of the sliding filament model, viz. zero length change of the

*Permanent address: Institute for Molecular Physics, University of Maryland, College Park, Maryland 20742.

filaments during contraction, may not be rigorously correct.[5]

In what follows there will be suggested an approach to the problem of contraction which, although based on an acceptance of the actin-myosin interaction theory, is derived primarily from structural considerations. Such a method circumvents the need for assumptions concerning the exact nature of the interaction but it does not preclude their use whenever unambiguous. No detailed description will be given here of either the structural or the biochemical aspects of muscle and its contraction; readers concerned with such considerations will find the recent review of Squire extremely informative.[6]

In this report a technical formulation of the problem appears in which the mathematical and theoretical aspects of the problem take precedence over the physiological. The work to be described draws heavily on the theses of T. W. Houk, Jr. and A. K. West, with whom will be published elsewhere a version of these ideas suited more properly to an audience concerned with both the biochemical and physiological aspects of vertebrate striated muscle and its contraction.[7]

Vertebrate striated muscle exhibits a hierarchial structure. Attached by tendons at either end to skeletal elements, the fibers of muscle are composed of fibrils 1-2μ in diameter. These in turn exhibit along their extended axes a repeating pattern of striations which defines the elementary, repeating unit of muscle - the sarcomere. Within the sarcomere is a centrally located array of filaments of the protein myosin. It, in turn, consists of a longitudinal, rod-like segment which retains the axial symmetry of the sarcomere and a second, two component, segment which lies at an acute angle with respect to the axis and with a polarity of direction which reverses at the median line of the sarcomere (M-line). The axial component is known as light meromyosin (LMM), while the heavier component (HMM) has both a rod-like portion (HMM-2) and a duplex, bulbous end (HMM-1) which contains the enzymatic sites for ATP dephosphorylation, one to each bulb. In cross-section the HMM-2 are hexagonal close packed. Also arrayed with the same symmetry is a double system of filaments of the protein actin, one system attached at either end of the sarcomere (Z-line) and interdigitating with the myosin system, the amount of overlap changing during contraction from its minimum value when the muscle is at rest to its maximum at maximum tension. The actin system has two other associated proteins which doubtless play a regulatory role: troponin and tropomyosin, the latter lying along the grooves of the double helix formed by the two elementary strands of associated monomeric actin subunits. In this sense actin resembles a double stranded pearl necklace, each pearl an actin monomer. The polarity (pitch) of the actin array reverses at the Z-line.

The motivation for a "structural model" is two-fold: In the first place, and in spite of some recent comments to the contrary,[8] the biochemical and structural properties of the contractile apparatus are not completely understood. Most of the present-day theories based on actin-myosin interaction prescribe intimate associations between the known biochemical events, primarily the dephosphorylation of ATP, and the hypothesized mechanical steps involved in the contractile process. Since the biochemical events usually are studied in homogeneous systems one cannot be certain that identical relationships are maintained in the sarcomere. Furthermore, the biochemical sequence need not correspond directly to a mechanical sequence. By contrast, a model developed primarily on structural grounds provides a framework to which details can be added as discovered.

Second, if one assumes that actin-myosin binding is in some way connected to the contractile event, it is clear that the myosin molecule must attach and detach from the actin filament regardless of the nature of the force-generating phenomenon. Any contractile theory must eventually accomodate such ligations if it is to be considered complete. The version to be offered here has yet to be considered for contractile processes, apart, that is, from two earlier occasions when certain aspects were presented in a preliminary version.[7] Recently, however, Walz, in a paper devoted principally to myosin MgATPase activity, described a sequence of steps consistent with our model (including negative cooperativity) and suitable for a contractile mechanism.[9]

Certain stochastic processes in muscle have been treated by Hill,[10] and Carlson,[11] from a study of intensity fluctuations in the spectrum of coherent light scattered from striated muscle, has concluded that the fluctuations arise as a consequence of a fluctuating driving force. Such a fluctuating force is the fundamental ingredient of our model.

Regard the actin-myosin interaction as a stochastic process. Specifically, describe the "state" of a given S-1 subunit (one of the two HMM-1 duplex bulbs) of a myosin molecule as a time dependent random variable corresponding to whether an S-1 is either bound or not to a given actin monomer. In a fully general treatment these random variables would not be independent but related to one another by the specific structural properties of the system, e.g. which thick filament the S-1 is connected to, where the S-1 lies on this filament, which series of actin monomers are available for binding and, lastly, the fact that the two variables corresponding to a given myosin molecule cannot be independent. Even to describe such a generalized set of random variables and their relationships would be difficult; fortunately, however, the physical properties of the contractile system allow for considerable simplification.

Experimental studies indicate that for normal sarcomere lengths the force generated by the contractile units is linearly proportional to the number of crossbridges which overlap the actin filament. This suggests that even though it may not be true in the strictest sense, one can treat the system as if each crossbridge operated independently of all others. Assume that each crossbridge contains one myosin molecule. This reduces to two the number of pertinent random variables. These two variables are not independent (assuming the S-1 stay attached to the myosin molecule), allowing a final reduction of the system to a single variable. Since one is interested in the progress of this crossbridge relative to a given actin filament, one can eliminate a number of time- and position-dependent factors first by assuming that the S-1 ligate to a linear lattice of binding sites, and then to confine one's consideration to that interval of time during which the S-1 are interacting with this sequence of binding sites. For the purpose of simplification, two additional assumptions are convenient. First assume that both of the S-1 units of a myosin molecule can interact with the binding sites and, second, assume that during the interval considered at least one S-1 remains attached (the analysis can be effected without this latter assumption but the details are more complicated and do not yield significantly more descriptive results).

Given the above assumptions, both the random variable and its state space are readily described. Take the condition of the S-1 subunits as a two-component random variable the value of which is an ordered doublet which denotes the specific site to which each of the S-1 is individually bound. If one is unbound, the value 0 is assigned to that member of the doublet. The fact that individual members of the S-1 pair are not independent of one another is reflected by a restriction that only a subspace, defined by restricting the integer separation of the heads to no greater a distance than the actin subunit separation, is accessible to this variable. The time dependence of the system is then completely determined by the stochastic transition rates between allowable states.

The model just described can be characterized mathematically as a continuous-time Markov process on a discrete state space with temporally homogeneous transition rates.

DEVELOPMENT OF THE MASTER EQUATION

Label the binding sites along the lattice by an index j ($j = 1, 2, \ldots, J$) running from the first actin site towards the Z disc, near which the J-site is presumed to be located (formally, $J \to \infty$). The system can exist in one of four basic states: (a) One head bound, the other free (doubly degenerate); (b) Both heads bound, on adjacent sites; (c) Both heads bound, a vacant site separating them. An ordered pair of indices $(j, j' > j)$ identifies the states, which can be conveniently listed in the neighborhood of the site m in the following order: (m-1, m+1); (0, m+1); (m, m+1); (m, 0). The four states in the neighborhood of any other site are exactly analogous to these, so one can prepare a list of the states available to the system with a new index k, $k = 1, 2, \ldots, 4n$:

Index:	1	2	3	4	5	6 ...
State:	(0,2)	(1,2)	(1,0)	(1,3)	(0,3)	(2,3)

If one defines a step in the Markov process as either a binding or unbinding of a single head, the only allowed transitions are those between states k and k' where $k' = k \pm 1$. In addition, the rates are translationally invariant in the sense that $k \to k'$ and $k + 4n \to k' + 4n$ (n any integer) correspond to the identical physical process carried out at different sites. Two clear cut cases can be distinguished: In the "inchworm" mode the heads are constrained to maintain their relative order and cannot pass one another on the lattice of binding sites; in the "flip-flop" mode the heads are permitted to swivel about one another in a limited way, thus to make available more binding sites to the unbound head. For each of these modes transitions can be assigned rates as follows (the labels are arbitrary):

Transition (Inchworm)	Rate
$(0, k+2+n) \to (k+1+n, k+2+n)$	a
$(k+1+n, k+2+n) \to (0, k+2+n)$	b
$(k+1+n, k+2+n) \to (k+1+n, 0)$	c
$(k+1+n, 0) \to (k+1+n, k+2+n)$	d
$(k+1+n, 0) \to (k+1+n, k+3+n)$	e
$(k+n, k+2+n) \to (k+n, 0)$	f
$(k+n, k+2+n) \to (0, k+2+n)$	g
$(0, k+2+n) \to (k+n, k+2+n)$	h

Transition (Flip-Flop)	Rate
$(k+n, 0) \to (k-2+n, k+n)$	ℓ
$(k+n, 0) \to (0, k+n)$	m
$(k+n, 0) \to (k-1+n, k+n)$	n
$(0, k+n) \to (k+n, k+1+n)$	q
$(0, k+n) \to (k+n, 0)$	r
$(0, k+n) \to (k+n, k+2+n)$	s

Notice that for the latter mode only those transitions which represent reversal of heads or reversal followed by the binding of the unbound head have been included. Furthermore, the "flip-flop" modes are not unique as written, e.g. ℓ is "equivalent" to m followed by h $(m \times h)$, $n = m \times a$, $q = r \times d$, and $s = r \times e$. In the master equation the redundancy will be eliminated. Here the display of states is mainly didactic.

The rates just described are representative of suggested physical states and transitions among them; they do not necessarily correspond to well-defined chemical events. Nonetheless, the specification (in its fundamental version) implies that no further division of the states is required to completely specify the kinetic behavior of the system; furthermore, the states are in "internal equilibrium" in the sense that chemical events which do not result in a change of state occur much faster than a transition among states.

Let P_k designate the occupation probability for the k'th state. Typical equations for the time development of the various P_k are

$$\frac{\partial P_1}{\partial t} = - a\ P_1 + b\ P_2 - m\ P_7$$

$$\frac{\partial P_4}{\partial t} = e\ P_3 - (f + g)P_4 + h\ P_5$$

$$\frac{\partial P_7}{\partial t} = - r\ P_1 + c\ P_6 - (d + e)P_7 + f\ P_8.$$

All such equations can be assembled into a master equation

$$\frac{\partial \vec{P}}{\partial t} = - [L] \cdot \vec{P} \tag{1}$$

where [L] is the transition rate matrix

$$\begin{bmatrix}
a & -b & 0 & \cdot & \cdot & \cdot & m & \cdot & \cdot & \cdot & 0 & \cdot\ \cdot \\
-a & (b+c) & -d & \cdot & \cdot & \cdot & 0 & \cdot & \cdot & \cdot & 0 & \cdot \\
0 & -c & (d+e) & -f & \cdot & \cdot & 0 & \cdot & \cdot & \cdot & 0 & \cdot \\
0 & 0 & -e & (f+g) & -h & \cdot & 0 & \cdot & \cdot & \cdot & 0 & \cdot \\
0 & \cdot & 0 & -g & (h+a) & -b & 0 & \cdot & \cdot & \cdot & m & \cdot \\
0 & \cdot & \cdot & 0 & -a & (b+c) & -d & \cdot & \cdot & \cdot & 0 & \cdot \\
r & \cdot & \cdot & \cdot & 0 & -c & (d+e) & -f & \cdot & \cdot & 0 & \cdot \\
0 & \cdot & \cdot & \cdot & 0 & 0 & -e & (f+g) & -h & \cdot & 0 & \cdot \\
0 & \cdot & \cdot & \cdot & 0 & \cdot & 0 & -g & (h+a) & -b & 0 & \cdot \\
\cdot & \cdot & \cdot & \cdot & 0 & \cdot & \cdot & 0 & -a & (b+c) & -d & \cdot \\
\cdot & \cdot & \cdot & \cdot & r & \cdot & \cdot & \cdot & 0 & -c & (d+e) & -f \\
\cdot & \cdot & \cdot & \cdot & 0 & \cdot & \cdot & \cdot & \cdot & 0 & -e & (f+g) \\
 & & & & & & & & & & & \cdot
\end{bmatrix} \tag{2}$$

and the state vector $\vec{P} \equiv (P_1, P_2, \ldots, P_k, \ldots)$. In what follows immediately only the "inchworm" mode will be considered, thus $m = r = 0$.

It was earlier mentioned that the myosin binding subunits (HMM) exhibited a polarity of direction reversing at the M line. Such a directionality of binding can be incorporated into this theory by requiring that $h = 0$, thus eliminating a step which would lead to a reversal of motion along the lattice - in fact, the most strongly backwards step - into the neighborhood of a site previously occupied. Now only the states 1 through 5 need be considered, for upon reaching 5 past states become irrelevant and 5 becomes, effectively, an initial state, 1'. The relevant master equation becomes

$$\frac{\partial}{\partial t} \begin{bmatrix} P_1 \\ P_2 \\ P_3 \\ P_4 \\ P_5 \end{bmatrix} = - \begin{bmatrix} a & -b & 0 & 0 & 0 \\ -a & (b+c) & -d & 0 & 0 \\ 0 & -c & (d+e) & -f & 0 \\ 0 & 0 & -e & (f+g) & 0 \\ 0 & 0 & 0 & -g & a \end{bmatrix} \cdot \begin{bmatrix} P_1 \\ P_2 \\ P_3 \\ P_4 \\ P_5 \end{bmatrix} \qquad (3)$$

SOLUTION OF THE REDUCED MASTER EQUATION

A convenient procedure for the solution of the master equation employs Laplace transforms. As the procedure is both straightforward and not especially revealing, in what follows a sketch only of the procedure will be presented. In vector form one has

$$\hat{\vec{P}}\,(s) = \int_0^\infty e^{-st}\,[I] \cdot \vec{P}\,(t)\,dt, \qquad (4)$$

and employing the identity

$$\int_0^\infty \frac{d}{dt}\hat{F}\,(t)\,e^{-st}\,dt = s\hat{F}(s) - F\,(o) \qquad (5)$$

one obtains the transformed master equation

$$s \cdot [I] \cdot \hat{\vec{P}}\,(s) - \vec{P}\,(o) = - [L] \cdot \hat{\vec{P}}\,(s), \qquad (6)$$

or

$$(s \cdot [I] + [L]) \cdot \hat{\vec{P}}\,(s) = \vec{P}\,(o), \qquad (7)$$

the system of equations requiring solution. Employing Cramer's rule, one obtains from the characteristic determinant a 5th order polynomial

$$\Delta = s^5 + \alpha\, s^4 + \beta\, s^3 + \gamma\, s^2 + \delta\, s \qquad (8)$$

where α, β, γ, δ are constants depending only on the transition rates: α, for example, is the sum (sign reversed) of the diagonal elements of [L]. Making use of Δ one obtains for the transformed probabilities

$$\hat{P}_k(s) = \frac{C_k^1 s^4 + C_k^2 s^3 + C_k^3 s^2 + C_k^4 s + C_k^5}{\Delta} \tag{9}$$

where the C_k^n are constants dependent upon the initial probabilities, $P_k(o)$, and upon the transition rates. For $k = 1 - 4$, $C_k^5 = 0$; thus $\hat{P}_{1-4}(s)$ are the quotient of a 3rd order polynomial by a 4th order. $\hat{P}_5(s)$, on the other hand, is one order higher in s and the numerator contains the non-zero constant term C_5^5.

To invert the transforms one needs a knowledge of the roots of Δ. Setting $\Delta = 0$ one finds immediately that $s = 0$ is a root. Only $P_5(t)$ will be affected by this fact, a constant term arising upon inversion from the presence of C_5^5. This assures that the probability of the absorbing state approaches unity for infinite time. The remaining problem is to find the roots of $s^{-1}\Delta = 0$, (essentially the eigenvalue equation for this reduced system) which can be accomplished with only minor inconvenience. Three possible situations occur: all four roots real; one pair complex, one pair real; two pair of complex (conjugate) roots. Generally in stochastic problems only the case corresponding to four real roots would be considered since detailed balance could be involved to assure real eigenvalues for the transition rate matrix.[12] Such is not the case here and the following will demonstrate the method of solution when complex roots occur.

Since the coefficients of the polynomial Δ are real, the roots are of the form:

$$s_1 = -A + iB \qquad s_3 = -C + iD$$
$$s_2 = -A - iB = s_1^* \qquad s_4 = -C - iD = s_3^* \tag{10}$$

which leads to the equation

$$\Delta = (s^2 + 2As + A^2 + B^2)(s^2 + 2Cs + C^2 + D^2)s \tag{11}$$

The method of partial fractions can be used next to reduce the $\hat{P}_k(s)$:

$$\hat{P}_k(s) = \frac{\xi_k s + \eta_k}{s^2 + 2As + A^2 + B^2} + \frac{\rho_k s + \lambda_k}{s^2 + 2Cs + C^2 + D^2} \quad (k=1, \ldots, 4) \tag{12}$$

$$\hat{P}_5(s) = \frac{\xi' s + \eta'}{s^2 + 2As + A^2 + B^2} + \frac{\rho' s + \lambda'}{s^2 + 2Cs + C^2 + D^2} + \frac{\theta}{s} \quad (13)$$

where the constants ξ_k, η_k, ... λ', θ depend upon the C_k^n and both the real and imaginary parts of the roots. Inversion of the Laplace transforms leads to the following expressions for the probabilities

$$P_k(t) = \frac{(\xi_k A + \eta_k)}{B} e^{-At} \sin Bt + e^{-At} \cos Bt + \quad (14)$$

$$\frac{(\rho_k C + \lambda_k)}{D} e^{-Ct} \sin Dt + e^{-Ct} \cos Dt \; (k = 1, \ldots, 4)$$

$$P_5(t) = \frac{(-\xi' A + \eta')}{B} e^{-At} \sin Bt + e^{-At} \cos Bt + \quad (15)$$

$$\frac{(-\rho' C + \lambda')}{D} e^{-Ct} \sin Dt + e^{-Ct} \cos Dt + \theta.$$

Thus the solutions exhibit damped periodic motion. If but one pair of complex roots exits, then the solution will continue to exhibit a periodic part together with simple exponential decays arising from the real roots. When all four roots are real, the solutions are pure exponential decay; they may oscillate but only aperiodically.[13]

THE MEAN FIRST PASSAGE TIME

A particularly important result which can be derived for this model of contraction is the mean first passage time (MFPT), corresponding to the time required for the system to reach state 5 from a given initial distribution $P_k(o)$. We begin by discussing the "inchworm" mode exclusively.

If the system is initially entirely in state 1 ($P_k(o) = \delta_{k1}$), then the MFPT corresponds to the time needed for a crossbridge to move from one binding site to the next, i.e. a unit step along the lattice, in the absence of an external load, the so-called zero load contraction velocity. The effects of a non-zero load will be described later.

The MFPT will be obtained following a prescription given by Weiss.[14] Let $P_v(t)$ be the probability that the system is any one of its states 1-4 at time t, then

$$P_v(t) = \sum_{k=1}^{4} P_k(t). \quad (16)$$

Also, let $\eta(t)$ be the <u>probability density</u> for the MFPT such that $\eta(t)dt$ represents the probability that the MFPT, $<t(1)>$, from state 1 to 5 lies in the interval $t<(<t(1)>)<t + dt$. By conservation,

$$P_v(t) = \eta(t)dt + P_v(t + dt), \tag{17}$$

or

$$\eta(t) = \frac{P_v(t) - P_v(t + dt)}{dt} \tag{18}$$

$$= - \frac{\partial P_v(t)}{\partial t}. \tag{19}$$

With the help of these equivalences one can define the MFPT as

$$<t(1)> \equiv \int_0^\infty t\eta(t)dt = \int_0^\infty P_v(t)dt. \tag{20}$$

The MFPT can also be written as

$$<t(1)> = \int_0^\infty [\sum_{k=1}^{4} P_k(t)]dt \tag{21}$$

and upon introduction of the Laplace transforms of $P_k(t)$, further simplified to

$$<t(1)> = \sum_{k=1}^{4} \hat{P}_k(o). \tag{22}$$

The required $\hat{P}(o)$ can be obtained by first setting $s = 0$ in Eq. (7) and then by multiplication with the inverse of [L]:

$$\hat{P}(o) = [L]^{-1} P(o). \tag{23}$$

But, as the initial state was chosen to be $P_k(o) = \delta_{k1}$, then

$$<t(1)> = \sum_{k=1}^{4} \sum_{j=1}^{5} ([L]^{-1})_{kj} \delta_{j1}$$

$$= \sum_{k=1}^{4} ([L]^{-1})_{k1}. \tag{24}$$

The result, after a straightforward yet tedious development of $[L]^{-1}$, is

$$\langle t(1)\rangle = \frac{1}{a}\{1 + \frac{b}{c}[1 + \frac{d}{e}(1 + \frac{f}{g})]\} + \frac{1}{c}[1 + \frac{d}{e}(1 + \frac{f}{g})] + \frac{1}{e}(1 + \frac{f}{g}) + \frac{1}{g}\,. \tag{25}$$

The usefulness of this result depends upon an estimation of the specific rates a through g. One helpful suggestion made and utilized for other studies by Podolsky [15] and by A. F. Huxley [16] consists of asserting that one (or a few) of the rates is sufficiently much slower than the rest to be rate-limiting.

Among the steps suggested for the "inchworm" mode, that one involving detachment of the leading head when both are adjacently bound is a prime candidate for the slowest step, u; therefore, set $c = u$. There are plausible reasons, originating in the biochemistry of contraction, for suggesting inequalities among the rates; for example, $a>h$, $f>c$, $e\simeq d$, $g>b$. [17] Furthermore, it is convenient to scale each of the rates with respect to u. A consistent set of rates is not difficult to achieve. [17] In the expression for $\langle t(1)\rangle$ one of the rates is absent; which one, dependent on the arbitrary choice of an initial state. Since the MFPT should be essentially independent of such a choice, it serves as a valuable criterion for the assignment of the individually scaled rates. One such set is

$$a = \frac{2}{3}\cdot 10u \qquad e = \frac{2}{3}\cdot 10u$$

$$b = \frac{3}{2}u \qquad f = \frac{3}{2}u$$

$$c = u \qquad g = \frac{9}{4}u$$

$$d = 10u \qquad h = \frac{4}{9}\cdot 10u$$

For such a choice the MFPT becomes

$$\langle t(1)\rangle = 5.2u^{-1} \tag{26}$$

Although the value of the coefficient changes somewhat from choice to choice, the changes are not significant and do not change the essential and plausible physical result that MFPT $\sim u^{-1}$.

To better appreciate the stochastic nature of the process, compare the minimum time required for the system to pass through its set of states directly with Eq. (26):

$$t_{min} = a^{-1} + c^{-1} + e^{-1} + g^{-1} \tag{27}$$

corresponding to a velocity three times as rapid.

INCLUSION OF THE FLIP-FLOP MODE: SOLUTION BY PERTURBATION

The procedure just described can be generalized to include the "flip-flop" dynamics by now setting h = 0 in the appropriate elements of [L], viz. the (8,9) and (9,9) positions. The system is again finite (9 states) and the transition rate matrix becomes

$$[L] = \begin{bmatrix} a & -b & 0 & 0 & 0 & 0 & m & 0 & 0 \\ -a & (b+c) & -d & 0 & 0 & 0 & 0 & 0 & 0 \\ 0 & -c & (d+e) & -f & 0 & . & . & . & . \\ 0 & 0 & -e & (f+g) & -h & 0 & . & . & . \\ 0 & . & 0 & -g & (h+a) & -b & . & . & . \\ 0 & . & . & 0 & -a & (b+c) & -d & . & . \\ r & . & . & . & 0 & -c & (d+e) & -f & . \\ 0 & . & . & . & . & 0 & -e & (f+g) & 0 \\ 0 & . & . & . & . & . & 0 & -g & a \end{bmatrix} \tag{28}$$

The MFPT becomes

$$\langle t(1)\rangle = \sum_{k=1}^{8} ([L]^{-1})_{kl}, \tag{29}$$

c.f. Eq. (22). Now, however, it is no longer a simple matter to invert [L]. If one supposes that the elements m and r are small compared to those appearing along the otherwise tridiagonal matrix, then a perturbation procedure immediately suggests itself. Physically this situation corresponds to an inclusion of the "flip-flop" mode with a manifestly smaller probability than the "inchworm." To effect the former, the duplex bulb would require the equivalent of a demi-piroutte, and the concomitant structural demands on the HMM subunit would seem to indicate that such events would be less likely than the less elegant but entropically more favorable "inchworm" mode.

Accordingly, the transition rate matrix is decomposed into an unperturbed stump (the tridiagonal part of [L]), and a small perturbation Δ, which is everywhere zero except for the elements m and r. The inverse can be expanded in powers of Δ, and to first order one obtains

$$[L]^{-1} \simeq [L_o]^{-1} \left([1] + \Delta\, [L_o]^{-1}\right). \tag{30}$$

The formal inversion of $[L_o]$ is achieved by the method of cofactors, as for Eq. (25). Since Δ is zero nearly everywhere, the evaluation of the MFPT from the elements of $[L]^{-1}$ is simplified considerably, only the first column of $[L]^{-1}$ being required. Thus,

$$[L]^{-1}_{k\ell} = \sum_{i=1}^{9} [L_o]^{-1}_{ki} \left(\delta_{i1} - \sum_{j=1}^{9} \Delta_{ij}\, [L_o]^{-1}_{j\ell}\right) \tag{31}$$

but Δ_{ij} is non zero only for the elements Δ_{17} and Δ_{71}, therefore

$$\langle t(1)\rangle = \left(1 - \Delta_{17}\, [L_o]^{-1}_{71}\right) \sum_{i=1}^{8} [L_o]^{-1}_{i1} - \Delta_{71}\, [L_o]^{-1}_{11} \sum_{i=1}^{8} [L_o]^{-1}_{i7}\,. \tag{32}$$

In keeping with the earlier argument, one can also here set the rates m and r as m = Mu and r = Ru. Full substitution of the rates into Eq. (32) leads to an expression for $\langle t(1)\rangle$ which is lengthy and consists of three distinct terms the first of which is:

$$\langle t(1)\rangle = I + II + III$$

$$\begin{aligned} I = {} & \frac{1}{a}\left[1+ \frac{b}{c}\left(1+ \frac{d}{e}\left(1+ \frac{f}{g}\left(1+ \frac{h}{a}\left(1+ \frac{b}{c}\left(1+ \frac{d}{e}\left(1+ \frac{f}{g}\right)\right)\right)\right)\right)\right)\right] \\ & + \frac{1}{c}\left[1+ \frac{d}{e}\left(1+ \frac{f}{g}\left(1+ \frac{h}{a}\left(1+ \frac{b}{c}\left(1+ \frac{d}{e}\left(1+ \frac{f}{g}\right)\right)\right)\right)\right)\right] \\ & + \frac{1}{e}[\quad] + \frac{1}{g}[\quad] + \frac{1}{a}[\quad] + \frac{1}{c}[\quad] + \frac{1}{e}[\quad] + \frac{1}{g} \end{aligned} \tag{33}$$

where the terms not identified can be read from the matrix, Eq. (28), by the evident rule employed for the two explicitly displayed. The second term is

$$II = - I \times m \times \frac{1}{e}\left(1 + \frac{f}{g}\right) \tag{34}$$

while the last, a bit messier, reads

$$III = - I' \times r \times [(\frac{bdfhbd}{acegace} + \frac{dfhbd}{cegace} + \frac{fhbd}{egace} + \frac{hbd}{gace} + \frac{bd}{ace} + \frac{d}{ce} + \frac{1}{e}) \times (1 + \frac{f}{g}) + \frac{1}{g}] \quad (35)$$

where I' is the leading term only of I.

This formidable result can be considerably simplified by substitution of the scaled rates with the result that

$$<t(1)> = \mu<t(1)>_o - u^{-1} (\alpha M + \beta R) \quad (36)$$

in which $<t(1)>_o$ is the unperturbed MFPT, Eq. (26), u the characteristic rate and the quantities α, β and μ are numerical constants. The values of α, β will typically lie in the interval 1-10 (determined in detail by the scaling assignment), while one expects on general grounds that μ be approximately 4, reflecting the fact that for such a stochastic process as envisioned here the time required for a crossbridge to travel two sites, instead of one, ought to be roughly proportional to the square of two - in general, to the square of the distance travelled.

The effect of the "flip-flop" mode is to decrease the MFPT since such transitions permit a crossbridge to take a two-site step. Yet, the effect is small, as the individual rates were so specified, and, naturally, linear in a first-order theory.

It is not obvious how to assign weight to the two modes considered. The "inchworm" may be the dominant mode since it corresponds to a sequence of short random walks rather than a single large one. One possible test of this idea rests in an estimation of the rate limiting step u, the so-called cycle rate. Given the distance between actin binding sites (one or a few actin monomers) and a value of an observed zero-load contraction velocity, it would be possible from either $<t(1)>_o$ or $<t(1)>$ to estimate u. The value obtained from $<t(1)>_o$ would necessarily be a lower bound. Such estimates in conjunction with an accurate measurement of u could aid in characterizing more precisely the detailed dynamics of contraction.

LOAD DEPENDENT TRANSITION RATES: HILL'S EQUATION

The literature of muscle contains an important phenomenological result due to A. V. Hill [18] which relates the velocity of contraction to the load on the muscle, P, relative to the maximum load, P_o, corresponding to zero velocity. It reads

$$(P + A)v = (P_o - P)B \quad (37)$$

where v is the velocity of shortening and A, B are constants. In the work so-far described only the zero load behavior of contracting muscle has been treated; nonetheless the model can be employed to examine the effects of load on the velocity and, indeed, it will be shown to agree with Hill's equation for small loads.

Let the actual load on the muscle P be small compared to P_o: $\lambda = (P/P_o) < 1$. Suppose also that the effect of load on a myosin bulb is smaller when the binding or unbinding occurs adjacent to the fixed head than when it occurs with an intervening site, and that the effect is to make it more difficult for the leading head, and easier for the trailing head, to bind. Finally, assume that if a particular attachment rate is increased by a fixed amount, then the corresponding detachment rate will be decreased in the same proportion and conversely. Whenever $\lambda \ll 1$ it is possible to imagine that the rates can be expanded in λ and terms only $O(\lambda)$ retained. For the rates e - h an additional increment/decrement is included by a coefficient n ($n > 1$) to reflect the assumption concerning intervening sites. Thus, the rates can be represented as

$a(1 + \lambda)$	$b(1 - \lambda)$
$c(1 + \lambda)$	$d(1 - \lambda)$
$e(1 - n\lambda)$	$f(1 + n\lambda)$
$g(1 - n\lambda)$	$h(1 + n\lambda)$

It has been demonstrated by Hill [19] that the constant $A \simeq P_o/4$. Together with the assumption that $\lambda \ll 1$, this fact permits Eq. (37) to be written as

$$\frac{1}{v} \simeq \frac{1}{4B}(1 + 5\lambda). \tag{38}$$

The maximum velocity v_o corresponds to $P = 0$ and in fact equals $P_o(B/A)$, which is approximately $4B$; thus

$$\frac{v}{v_o} \simeq 1 - 5\lambda \tag{39}$$

a particularly compact version of Hill's equation for small λ. Substitution of the load-dependent rates into Eq. (25) for the MFPT ("inchworm" exclusively) results in

$$\langle t(1)\rangle_\lambda - u^{-1}\,(5.2 - 10.5\lambda + 9.0n\lambda). \tag{40}$$

Noting that $v_o \sim \langle t(1)\rangle_o^{-1}$ and $v \sim \langle t(1)\rangle_\lambda^{-1}$, one obtains after a bit of algebra

$$\frac{v}{v_o} \simeq 1 + 2.0\lambda - 1.7n\lambda \qquad (41)$$

a result in agreement with Eq. (39) for $n \simeq 4$, suggesting a reasonable estimate for the effect of an intervening site on the rates of attachment and detachment.

This particularly simple result is by no means trivial. Only a few plausible assumptions are required to establish the equivalence between the stochastic model and Hill's equation. In its simplest version the stochastic model is consistent with experimental observations at small loads. This fact raises the distinct possibility that at some future point, when more is known about the contraction of muscle at the molecular level, it may be possible to refine and extend the present model to agree with experimental data over the full range of loads.

THE LENGTH TENSION ISOTHERM

Over a considerable range of actin-myosin overlap the maximum tension which a muscle develops is proportional to the extent of overlap. From the point of view of the model discussed here the explanation for this observation is direct and simple; for, if maximum tension results from some fraction (perhaps unity) of all crossbridges being bound, and if each crossbridge makes an essentially independent contribution to the tension developed, then tension $\sim$ extent of binding $\sim$ fraction of overlap.

When all of the crossbridges are in register the tension levels off and remains so, only to decrease when the sarcomere lengths become very short. The decrease may be accounted for by supposing that physical deformation of the filaments, occuring when the muscle approaches maximum shortening, forces the crossbridges to such linear densities that interference results, and an increasingly smaller fraction of the heads remains in register with actin binding sites.

Such an explanation, of course, is a simple, natural consequence of the sliding filament model and is not unique to this discussion. Simplicity notwithstanding, there are two important parameters which can be estimated from the model and which are not as yet directly available from experiment: the force generated per crossbridge, and the energy release necessary for transit from site to site. A rough estimate of the force per crossbridge can be made as follows:

$$F = P_a/\sigma N \qquad (42)$$

where the force is taken per crossbridge at maximum (full) overlap, P_a is the tension-area^{-1} generated by a muscle fiber at maximum overlap, σ is the cross-sectional density of myosin filaments, and N is the number of crossbridges per myosin filament in each half-sarcomere. Approximate values for these parameters are (references in parenthesis):

$$P_a \simeq 1 - 5 \times 10^5 \text{ dynes cm}^{-2} (20), \; \sigma \simeq 7 \times 10^{10} \text{ cm}^{-2} (21),$$

$$N \simeq 180\text{-}360 \; (18)$$

from which a value of F is obtained as $F \simeq (0.4 - 4) \times 10^{-7}$ dynes. T. Hill and White, using similar considerations, have suggested a value for F of $\simeq 2 \times 10^{-7}$ dynes.[22]

Actin binding sites cannot be significantly less than 100 Å apart, thus a lower bound for the work done by a single crossbridge in moving one site is 10^{-13} erg. On the other hand, if ATPase activity occurs upon detachment, and if detachment involves at least a pair of steps for motion from site to site, each step releasing 5 k cal/mole,[23] then the available energy for contraction is of the order of 10 k cal mole^{-1}. From the value of F deduced above one obtains a lower bound estimate of about 3 k cal mole^{-1}, numbers in surprisingly good agreement considering the nature of the approximations involved.

Such estimates as appear here are predicated on the assumption that the force per crossbridge is constant so that actual tension is determined by only the extent of overlap and the fraction of bound heads. No important change in result would occur if the force per crossbridge were variable while the bound fraction were constant.

ACKNOWLEDGEMENTS

For the work described here I owe particular debts to T. William Houk and Arthur K. West who with considerable patience continue to instruct me on the matter of muscle. We three owe a collective debt to V. Adrian Parsegian for trying to keep us scrupulously honest.

REFERENCES

1. J. Hanson and H. E. Huxley, Nature 172, 530 (1953); A. F. Huxley and R. Niedergerke, ibid. 173, 971 (1954).
2. R. E. Davies, Nature 199, 1068 (1963); T. L. Hill, Proc. Nat. Acad. Sci. USA 59, 1194 (1968); A. G. Lowery, J. Theoret. Biol. 20, 164 (1968); C. F. W. McClare, ibid. 35, 569 (1972); V. I. Deshcherevskii, Biofizika 13, 928 (1968); H. E. Huxley, Science 164, 1356 (1969); M. V. Volkenstein, Biochem. Biophys. Acta 180, 562 (1969); A. F. Huxley, Proc. Roy. Soc. London B183, 83 (1973).
3. W. C. Ullrick, J. Theoret. Biol. 15, 53 (1967); D. T. Warner, ibid. 26, 289 (1970); C. T. Dragomir, ibid. 27, 343 (1970); D. B. Shear, ibid. 28, 531 (1970); G. F. Elliot, L. M. Rome and M. Spencer, Nature 226, 417 (1970); L. C. Yu, R. M. Dowben and K. Kornacker, Proc. Nat. Acad. Sci. USA 66, 1199, (1970).

4. L. Mandelkern and B. Van Duzee, Biochem. and Biophys. Res. Comm. 51, 1048 (1973).
5. J. E. Berger, L. Herman and P. Dreizen, Biophys. Soc. Absts. 13, 183a (1973).
6. J. M. Squire, Ann. Rev. Biophysics and Bioengineering 4, 137 (1975).
7. A. G. De Rocco and T. W. Houk, Technical Report No. 71-070, Dept. of Physics and Astronomy, Univ. of Md., June, 1970; A. G. De Rocco and T. W. Houk, Biophys. Soc. Absts. 13, 325a (1973); T. W. Houk, Thesis (Ph. D., physics), Univ. of Md. (1971); A. K. West, Thesis (Ph. D., physics), Univ. of Md. (1973).
8. C. F. W. McClare, Nature 240, 88 (1972).
9. F. G. Walz, J. Theoret. Biol. 41, 357 (1973).
10. T. L. Hill, Prog. Biophys. Mol. Biol. 28, 267 (1974); T. L. Hill, private communication.
11. F. D. Carlson, Biophys. J. 15, 633 (1975).
12. I. Oppenheim, K. E. Shuler and G. H. Weiss, Adv. Molec. Relax. Proc. 1, 817 (1967).
13. K. E. Shuler, Phys. Fluids 2, 442 (1959).
14. G. H. Weiss, Adv. Chem. Phys. 13, 1 (1966).
15. R. J. Podolsky, Structure and Function of Muscle, 2, 359 (1960).
16. A. F. Huxley, Prog. Biophys. Biophys. Chem. 7, 255 (1957).
17. A. K. West, Thesis (Ph.D., physics), Univ. of Md, College Park, MD. (1973), Appendix B.
18. J. R. Bendall, "Muscles, Molecules and Movement" (American Elsevier, New York, 1969).
19. A. V. Hill, "Trails and Trials in Physiology" (Williams and Wilkins, Baltimore, 1966).
20. R. J. Ashley, J. Theoret. Biol. 36, 339 (1972).
21. H. E. Huxley, Science 164, 1356 (1969).
22. T. L. Hill and G. M. White, Proc. Nat. Acad. Sci. USA 61, 514 (1968).
23. R. J. Podolsky and M. F. Morales, J. Biol. Chem. 218, 945 (1956).

Dr. De Rocco is a Professor in the Institute for Molecular Physics at the University of Maryland and has been a member of the Physical Sciences Lab., Division of Computer Research and Technology of the National Institutes of Health. Recently he has been working on the theory of fluids and models for biorhythms and muscle behavior. He first met Julius in Washington, D.C.

THE RENORMALIZATION GROUP AND THE UNIVERSALITY OF CRITICAL EXPONENTS

Melville S. Green and Prabodh Shukla
Temple University, Philadelphia, Pa. 19122

ABSTRACT

A class of renormalization groups characterized as the general Gaussian renormalization group is defined. It includes the sharp cutoff renormalization group of Wilson, the linear renormalization group of Bell and Wilson, and the incomplete-integration renormalization group of Wilson. The renormalization group approach to critical phenomena and the expansion around four dimensions is reviewed. The fixed point of the general Gaussian renormalization group is determined to order ε^2, where $\varepsilon = 4 - d$ (d is the dimensionality of the system), and the critical exponents are determined to the same order. The main conclusion is that the exponents do not depend on the cutoff functions of the renormalization group.

INTRODUCTION

One of the most beautiful but puzzling features of the plethora of phenomena in the neighborhood of critical points[1-5] and second-order phase transitions is their universality. With appropriate redefinition and rescaling, the measured relationships among physical variables are universal for a wide variety of systems and substances. Thus changes in the size and shape of intermolecular potential functions change the thermodynamic functions near the critical point only by trivial scale factors.[6] A theoretical basis for the understanding of universality has been given by Kadanoff[4] in the concept of relevant and irrelevant perturbations in the Hamiltonian describing a physical system. Certain changes in the Hamiltonian such as those describing the range and shape of intermolecular forces are irrelevant and produce no essential qualitative change in the phenomena. On the other hand, changes in discrete parameters such as the dimension of space or of the dimension of the order parameter produce essential qualitative change. Thus fluids and magnets which are qualitatively similar among themselves but differ in the dimension of the order parameter are qualitatively different from each other.[7]

The renormalization group theory has been very successful both in explaining general qualitative features of critical phenomena as well as in determining quantitative features such as exponents and scaling functions.[8,9] In this theory the Hamiltonian describing a system is transformed into a series of Hamiltonians in some sense equivalent to it through the operations of a group called the renormalization group. A critical point is associated with a fixed point of this group, i.e., a Hamiltonian which is unchanged by the operations of the group. Mathematically, the operation of renormalization is a transformation on the Hamiltonian which preserves spatial uniformity and leaves the total free energy unchanged.[10] Physically, the operation of renormalization consists in the elimination of fluctuations of length scale smaller than some cutoff combined with a rescaling of all distances in the Hamiltonian.

It is clear that a renormalization transformation can be realized in not one but many, in fact a continuity of ways. How then, can a theory which involves an "ad libitum" element, lead to a unique and universal behavior in the neighborhood of the critical point? It is to this question that the present paper addresses itself.

The concept of renormalization group is a very general one and we do not attempt to show independence of physical results of the general renormalization group. We define, rather, a class of renormalization groups of which the sharp cutoff group of Wilson (recursion relation)[8], a smooth cutoff group proposed by Wilson and Kogut[9] and the linear group proposed by Bell and Wilson[11] are special or limiting cases. We call this class the general Gaussian renormalization group. We confine ourselves also to the case in which the order parameter is one-dimensional.

The space dimension four is special in the theory of critical phenomena and the renormalization group in that two fixed points, the so-called Gaussian fixed point and the non-Gaussian fixed point which are distinct in two and three dimensions coalesce in four dimensions. This fact makes the case of four dimensions explicitly solvable and provides the basis for one of the most successful methods for obtaining explicit results from the renormalization group approach. In this method an expansion is made in the parameter $\varepsilon = 4 - d$ where d is the spatial dimension.[12] This parameter is fantasized as a continuous variable but the results are obtained by setting ε equal to integer values corresponding to integer dimensions. We compute the exponent η to order ε^2, the first non-vanishing order, and show that it is independent of the two parameter functions which define a particular Gaussian renormalization group. We also determine the other thermodynamic exponents to order ε and show that they are independent of the parameter functions.

The outline is as follows. In section (2) we motivate the definition of the general Gaussian RNG, and derive a differential equation as the generator of this group. Our presentation of the "incomplete-integration" procedure is slightly different from the original of Wilson and Kogut[9], in particular, we present a somewhat more general group differential equation. In section (3) the RNG transformation is applied to a Gaussian Hamiltonian. A nontrivial Gaussian fixed point is discussed and the critical exponents are computed. The main purpose of this section is to explain the ideas of the RNG in a case relatively free from mathematical complications. In section (4) we consider a general Hamiltonian and motivate the epsilon expansion. We determine the general fixed-point Hamiltonian to order ε^2. In section (5) the relevant eigenvectors and their eigenvalues are calculated to order ε^2 and the critical exponents determined to the same order. A formula for the exponent η is obtained which, in order ε^2, appears to depend explicitly on the parameters of the renormalization group. However, under further analysis presented in section (6), the dependence of η on the parameters of the RNG is found to be only apparent. It is found that all critical exponents are universal independent of the cutoff function and other free parameters of the renormalization group.[13]

RENORMALIZATION GROUP TRANSFORMATION

2.1 General Characterization of a Renormalization Group Transformation

A renormalization transformation is one which changes the Hamiltonian of the system into another whose partition function yields the same free energy. If we denote the Hamiltonian of the system by H and the transformed Hamiltonian by H' then

$$\ell n \quad \text{trace} \quad \frac{H'}{kT} = \ell n \quad \text{trace} \quad \frac{H}{kT} \tag{2.1}$$

where k is the Boltzmann's constant and, T is the temperature. It will be convenient to combine the factor $(kT)^{-1}$ with the Hamiltonian and write

$$\ell n \quad \text{trace} \quad H' = \ell n \quad \text{trace} \quad H \tag{2.2}$$

where $H = \frac{1}{kT} H, \quad H' = \frac{1}{kT} H'$.

We define a transformation R by

$$H' = RH \tag{2.3}$$

A renormalization transformation is intended to produce a new Hamiltonian with the same qualitative properties as the original. Thus the new Hamiltonian will be spatially uniform and isotropic with short range intermolecular forces if the original Hamiltonian had these properties. An important qualitative feature of the Hamiltonian which need not be preserved in a renormalization transformation is the volume to which the system is confined. A scaling transformation which simply multiplies all distances among particles by a constant factor is a renormalization transformation which changes the volume of the system and any renormalization transformation can be considered to be composed of a renormalization transformation which does not change the volume and a scaling transformation.

We will deal with systems whose state can be represented by a scalar field $S(\vec{x})$. Such a system is the Ising model in which $S(\vec{x})$ can be thought of as the value of scalar spin variable at the lattice site $\vec{x}$. The fact that $S(\vec{x})$ takes on values only at discrete lattice points means that its Fourier components

$$S_{\vec{k}} = \frac{1}{V^{\frac{1}{2}}} \int_V S(\vec{x}) \exp i\vec{k}.\vec{x}\, d\vec{x}$$

are zero for $\vec{k}$ outside the first Brillouin zone of the lattice. We will consider a somewhat more general field such that the Fourier components of $S(\vec{x})$ approach zero very rapidly but do not necessarily vanish for $\vec{k}$'s greater than some cutoff k_c. The particles of the system will be separated by distances greater than $2\pi\, k_c^{-1}$. For a finite volume the wavevector $\vec{k}$ can take on only certain discrete values which are distributed in $\vec{k}$- space with density $(2\pi)^{-d}\, V$. The effective number of Fourier components is therefore finite and of the order of $(2\pi)^{-d}\, k_c^d\, V$. A renormalization transformation which does not change the volume will in general reduce the cutoff momentum to, say, $k_c' < k_c$, thereby increasing the effective minimum spacing of particles. The effective number of Fourier components will then be $(2\pi)^{-d}\, k_c'^d\, V < (2\pi)^{-d}\, k_c^d\, V$. A scale transformation which relabels each $S_{\vec{k}}$ with another wavevector $k' = \lambda\vec{k}$ does not change the effective number of Fourier components but will change their density in $\vec{k}$- space as well as the cutoff wavevector and the effective minimum spacing between particles. In particular it is possible to restore the minimum spacing to its original value by choosing $\lambda = k_c\, k_c'^{-1}$. The volume of the system will be reduced in the ratio λ^{-d}. A scale transformation will change the field $S(\vec{x})$ into a new field $S'(\vec{x}')$ such that $S'(\vec{x}') = S(\lambda\vec{x}')$ i.e. it will associate the value of the new field at $\vec{x}'$ with the old field at the larger distance $\lambda\vec{x}'$. The component of the new field at $\vec{k}'$ will be

$$S'_{\vec{k}'} = \frac{1}{V'^{\frac{1}{2}}} \int S'(\vec{x}') \exp i\vec{k}'.\vec{x}'\, d\vec{x}'$$

$$= \frac{1}{V^{\frac{1}{2}}}\, \lambda^{-\frac{d}{2}} \int S(\vec{x}) \exp i\vec{k}.\vec{x}\, d\vec{x}$$

$$= \lambda^{-\frac{d}{2}}\, S_{\vec{k}}$$

Thus for the Fourier components of the field a scale transformation will consist in a relabeling of the old field and multiplication by the factor $\lambda^{-\frac{d}{2}}$, while for the Hamiltonian the scale change will consist in the substitution of the old spins by their expressions in terms of the new

$$H'(S'_{\vec{k}'}) = H(S_{\vec{k}})$$

The resulting Hamiltonian will be a function of the same effective number of Fourier components whose wavevectors, however, are distributed less densely in a larger sphere. If an appropriate scale transformation is combined with a non-volume changing renormalization transformation the resulting Hamiltonian will be a function of Fourier components whose wavevectors are distributed in a sphere of the same size as the original Hamiltonian but with a reduced density. In coordinate space this means that the effective spacing between particles will remain unchanged but the size of the system will be reduced. It is thus conceivable that we may find a fixed point Hamiltonian i.e. a Hamiltonian whose properties are unchanged by a renormalization transformation.

$$H^* = R\,H^* \tag{2.4}$$

The equality sign in this equation refers to equality of intermolecular forces not, of course, identity of the volume to which the system is confined. Wilson[9] and Kadanoff[13] have shown how the behavior of thermodynamic functions and interparticle correlation functions near a critical point can be related to the behavior of the renormalization group near a fixed point. We will not pursue this aspect here but rather summarize what we have said about renormalization groups by the statement that a renormalization group which is likely to have a fixed point will consist of the combination of a non-volume changing renormalization transformation with a simple scale change.

We consider first the non-volume changing renormalization transformation. Most such transformations have been constructed from linear transformations on the Boltzmann factor

$$\exp H'(S') = \int_S T(S',S)\, \exp H(S)\, ds \tag{2.5}$$

where S' and S stand for the set of Fourier components of the new and old fields respectively. Since the set of wavevectors is infinite the integral $\int_S$ must be understood as a functional integral. We may assure that the transformation from $H \rightarrow H'$ is a renormalization by requiring that

$$\int_{S'} T(S',S)\, dS' = 1 \tag{2.6}$$

for then integrating both sides of equation (5) over S', assuming reversibility of the order of the functional integrals, yields

$$\begin{aligned} \int \exp H'(S')\, dS' &= \int_{S'} \int_{S} T(S',S)\, \exp H(S)\, dS dS' \\ &= \int_S \Big[\int_{S'} T(S',S)\, dS'\Big] \exp H(s)\, dS \\ &= \int_S \exp H(S)\, dS \end{aligned} \tag{2.7}$$

Thus the partition function and thermodynamic potential for both Hamiltonians are the same. It is natural also to assume that the elements of $T(S',S)$ are all positive in order that a positive Boltzmann factor yield a positive Boltzmann factor. Finally, we must assume that T operating on a spatially uniform Boltzmann factor produces a spatially uniform Boltzmann factor. This means that if we substitute $\exp(i\vec{k}.\vec{a})\, S_{\vec{k}}$ for $S_{\vec{k}}$ and $\exp(i\vec{k}'.\vec{a})\, S_{\vec{k}}'$, T remains unchanged for any choice of $\vec{a}$.

As a function of S', $T(S',S)$ has the properties of a normalized probability distribution. It is natural therefore to consider the case in which this distribution is Gaussian in S'. We will in fact consider the case in which T is Gaussian in both S' and S.

$$T(S',S) = \exp - Q(S',S)$$

where Q is a positive definite quadratic in S' and S. Such a form will in general contain terms of the form $S_{\vec{k}}S_{\vec{\ell}}$, $S'_{\vec{k}'}S'_{\vec{\ell}'}$, and $S'_{\vec{k}'}\, S_{\vec{k}}$. The spatial uniformity condition will require that only such terms appear for which $\vec{k} + \vec{\ell} = 0$, $\vec{k}' + \vec{\ell}' = 0$, and $\vec{k}' + \vec{k} = 0$. Thus we may write

$$T(S',S) = \exp - \sum_{\vec{k}} \Big(a(k)\, S'_{\vec{k}}S'_{-\vec{k}} + 2b(k)\, S'_{k}S_{-k} + c(k)\, S_{\vec{k}}S_{-\vec{k}}\Big)$$

T must be a product of factors, each one referring to a single value of $\vec{k}$. The requirement that $T(S',S)$ be normalized with respect to S' for all S requires that each factor be normalized with respect to $S'_{\vec{k}}$ for all $S_{\vec{k}}$. This means that we may write

$$T(S',S) = \exp - \sum_{\vec{k}} A(k)(S'_{\vec{k}} - B(k)S_{\vec{k}}) \times (S'_{-\vec{k}} - B(k)\, S_{-\vec{k}})$$

In order to preserve rotational symmetry $A(k)$, $B(k)$ are functions of the length of $\vec{k}$. In order that T should not generate long range forces $A(k)$ and $B(k)$ are analytic in k^2. The combination of this volume preserving transformation with a scale transformation is the general Gaussian renormalization transformation. We will present below the general Gaussian renormalization group in a differential form containing two parameter functions $\beta(k)$ and $\gamma(k)$. $\beta(k)$ corresponds to $B(k)$ while $\gamma(k)$ corresponds to $A(k)$. We note that the sharp cutoff renormalization group (Wilson's recursion relation)[9] can be understood as a limiting special case of general Gaussian renormalization transformation in which $B(k) \equiv B$

(independent of k), $A(k) = \infty$ for $k < \frac{1}{2}$, $A(k) = 0$ for $k > \frac{1}{2}$. For $k > \frac{1}{2}$ $T(S',S)$ simply integrates over $S_{\vec{k}}$ while for $k < \frac{1}{2}$ it sets $S_{\vec{k}} = S_{\vec{k}}$.

2.2 Wilson's Incomplete-Integration Renormalization Group Transformation.

We shall use Gaussian transformation kernels to construct a renormalization group transformation with the properties enunciated in the preceding section. The construction shall proceed in two stages. First, consider the transformation T' defined below

$$\exp H'[S'] = T' \exp H[S]$$

$$= C \int_{So} \exp \{-\tfrac{1}{2} \int_{\vec{k}} \gamma_1^{-1}(ke^t) |S_{\vec{k}} - \exp\{-\alpha_1(ke)\} S_{\vec{k}}|^2\}$$

$$\times \exp H[S] \qquad (2.8)$$

where $H[S]$ is the initial Hamiltonian which is transformed into the renormalized Hamiltonian $H'[S]$ by the transformation T'. A factor of $(k_B T)^{-1}$, where k_B is the Boltzman constant and T the temperature of the initial system, is implicit in each Hamiltonian. S denotes the set $\{S_{\vec{k}}\}$ of the Fourier component of a classical field S_x, $-\infty \le S_{\vec{x}} \le +\infty$, associated with the Hamiltonian H. Similarly S' denotes the set of Fourier components $\{S_{\vec{k}}\}$ associated with H. The wave-vector k is a d-dimensional vector of magnitude k. C appearing in equation (2.8) is a constant and $\gamma(k)$ and $\beta(k)$ are arbitrary analytic functions of k^2.

We work with Hamiltonians that can be defined in terms of an infinite set of arbitrary functions $u_2(k)$, $u_4(\vec{k}, \vec{k}_1, \vec{k}_2, \vec{k}_3)$,... etc. Further, following a common practice, we restrict ourselves to Hamiltonians which preserve the symmetry $S_{\vec{k}} \to - S_{\vec{k}}$ and write

$$H[S] = -\frac{1}{2} \int_{\vec{k}} u_2(\vec{k})\, S_{\vec{k}} S_{-\vec{k}}$$

$$- \frac{1}{4!} \int_{\vec{k}} \int_{\vec{k}_1} \int_{\vec{k}_2} u_4(\vec{k}, \ldots, \vec{k}_3)\, S_{\vec{k}} S_{\vec{k}_1} S_{\vec{k}} S_{\vec{k}_3}$$

$$- \frac{1}{6!} \int_{\vec{k}} \cdots \int_{\vec{k}_4} u_6(\vec{k}, \ldots, \vec{k}_5)\, S_{\vec{k}} S_{\vec{k}_1} S_{\vec{k}_2} S_{\vec{k}_3} S_{\vec{k}_4} S_{\vec{k}_5} \qquad (2.9)$$

where $\vec{k}_3 = -\vec{k} - \vec{k}_1 - \vec{k}_2$, $\vec{k}_5 = -\vec{k} - \vec{k}_1 - \vec{k}_2 - \vec{k}_3 - \vec{k}_4$,... etc. are chosen to ensure translational invariance.

The renormalization group transformation acts on the infinite dimensional space S formed by the functions $u_2(\vec{k})$, $u_4(\vec{k}, \vec{k}_1, \vec{k}_2, \vec{k}_3)$,... etc. The Hamiltonian expressed by the equation (2.9) is general and need not refer to any specific physical system. However, for convenience, we may assume that it refers to a ferromagnet. The set of variables $\{S_{\vec{k}}\}$ can then be understood as the Fourier components of the classical spin-field $S_{\vec{x}}$ and the functions u_2, u_4,... etc. can be understood to be the Fourier transforms of the two-spin and the four-spin interactions etc. in the coordinate space. The functions u_2, u_4,... etc. must be analytic in the Fourier space and must be cutoff for large values of the wave-vector. The requirement of analyticity in Fourier space is necessary to ensure that the interactions are local in the coordinate space.[14] The cutoff for large values of the wave-vector is necessary to be able to carry out the integration over S in equation (2.8) for the terms of H involving the products of four spins or more.

The transformation T', defined in equation (2.8), preserves the partition function of the initial Hamiltonian. To demonstrate this, we integrate over the spin variables S_k on both sides of the equation (2.8). This results in,

$$\int_{S'} \exp H'[S] = \text{constant} \times \int_{S} \exp H[S] \tag{2.10}$$

where the constant comes from performing the integral over the Gaussian transformation kernel. The value of this constant is independent of H[S] and has no effect on the calculation of any correlation functions for the Hamiltonian H[S]. We can, therefore, interpret equation (2.10) as the expression of the equivalence of the partition functions of the Hamiltonians H and H' respectively.

The transformation T', with proper choices of the functions $\gamma_1(ke^t)$ and $\alpha_1(ke^t)$, can integrate out the large wave-vector Fourier components $S_{\vec{k}}$. We assume that the fields $\{S_{\vec{k}}\}$, associated with the intial Hamiltonian H, are bounded in the sense that there exists a constant C and an increasing analytic function of k^2, $\beta(k)$, such that

$$C \exp\{-\beta(k)\} > S_{\vec{k}} \tag{2.11}$$

and choose

$$\alpha_1(ke^t = bt + \beta(ke^t) - \beta(k); \tag{2.12}$$

$$\gamma_1(ke^t) = [1 - \exp\{-2\alpha_1(ke^t)\}], \tag{2.13}$$

where b is a constant. We shall show below that the transformation T', with the choices of $\alpha_1(ke^t)$ and $\gamma_1(ke^t)$ expressed by equations (2.12) and (2.13), transforms the Hamiltonian H into a new Hamiltonian whose fields are bounded as follows.

$$C \exp\{-\beta(ke^{t})\} > S'_{\vec{k}} \tag{2.14}$$

Equation (2.14) implies that the cutoff function for the fields of the new Hamiltonian is the cutoff function for the fields of the old Hamiltonian shifted towards the smaller k value by an amount e^t. The shift in the cutoff is due to the fact that the transformation T' integrates out effectively an amount $(1-e^{-t})$ of the large wave-vector components $\{S_k\}$ of H.

Equation (2.13) is not the most general choice for the function $\gamma_1(k)$ to construct an effective cutoff e^{-t} on k. This choice has been made to derive Wilson's incomplete-integration renormalization group equation. Later a more general choice of $\gamma_1(k)$ will be made.

To see how T' achieves an effective cutoff e^{-t} on k associated with H consider the following simple choices for the parameters C and $\gamma_1(k)$

$$\gamma_1(k) \equiv \gamma_1(0), \tag{2.15}$$

$$C = \frac{1}{2\pi\gamma_1(0)} C' \tag{2.16}$$

where $\gamma_1(0)$ and C' are constants. Combining equations (2.8), (2.12), (2.15) and (2.16)

$$\exp H'[S'] = C' \frac{1}{2\pi\gamma_1(0)} \int_{S^o} \exp\{-\frac{1}{2}\gamma_1^{-1}(0)$$

$$\times \int_{\vec{k}} |S_{\vec{k}} - \exp\{-bt-\beta(ke^{t}) + \beta(k)\} S_{\vec{k}}|^2\} \exp H[S] \tag{2.17}$$

In order to understand the function of the transformation T' in its most simple form, we let $\gamma_1(0) \to 0$ in equation (2.17). In this case,

$$\exp H'[S']$$

$$= C' \int_S \delta(\int_{\vec{k}} \{S_{\vec{k}} - \exp\{-bt - \beta(ke^t) + \beta(k)\}S_{\vec{k}}\})$$

$$\times \exp H[S] \qquad (2.18)$$

The symbol δ appearing immediately after the integral sign in equation (2.18) denotes Dirac's delta function. It is seen from equation (2.18) that the effect of T' is to integrate out the variables $\{S_{\vec{k}}\}$ which are related to the old variables by the equation

$$S'_{\vec{k}} = \exp\{-bt - \beta(ke^t) + \beta(k)\}S_{\vec{k}} \qquad (2.19)$$

Equation (2.14), which we asserted earlier, follows from combining equations (2.11) and (2.19). The case when $\gamma_1(0)$ is not infinity constitutes an extension of the above analysis and the effect of T' is still to introduce an effective cutoff e^{-t} on the wave-vector k.

For a continuous variable parameter t, the transformation defined by equations (2.8), (2.12) and (2.13) can be equivalently expressed by the following differential equation for the functional $\exp H'(S')$

$$\frac{\delta}{\delta\alpha'_t(ke^t)} \exp H'(S') = \frac{\delta}{\delta S_{\vec{k}}}\left[\frac{\delta}{\delta S_{-\vec{k}}} + S'_{\vec{k}}\right] \exp H'[S'] \qquad (2.20)$$

The equivalence of the equation (2.20) with the equation (2.8), (2.12) and (2.13) can be verified by substitution. The functional differential equation for $H'[S']$ is

$$\frac{\partial H'}{\partial t} = \int_{\vec{k}} \frac{\partial\alpha'_t(ke^t)}{\partial t}$$

$$\times \left[\frac{\delta H'}{\delta S'_{\vec{k}}} \frac{\delta H'}{\delta S'_{-\vec{k}}} + \frac{\delta^2 H'}{\delta S'_{\vec{k}}\delta S'_{-\vec{k}}} + S'_{\vec{k}} \frac{\delta H'}{\delta S'_{\vec{k}}}\right] \qquad (2.21)$$

The transformation defined by equation (2.21) does not generate a group since the right-hand-side of equation (2.21) explicitly depends upon the parameter t. The removal of this "deficiency" constitutes the second stage in the construction of the renormalization group transformation. We scale the wave-vector and the field as follows[9]

$$K' = Ke^t \qquad (2.22)$$

$$S'_{\vec{k}'} = \exp\{-\frac{1}{2} dt\} S'_{\vec{k}} \tag{2.23}$$

The derivative $\frac{\partial H'}{\partial t}$ in equation (2.21) is for a fixed function S'_k. It is related to the derivative $\frac{\partial H}{\partial t}$ for fixed S'_k, by the following relation

$$\frac{\partial H'}{\partial t}\Big|_{S'_{\vec{k}}} = \frac{\partial H'}{\partial t}\Big|_{S'_{\vec{k}'}} + \int_{\vec{k}'} \frac{dS'_{\vec{k}'}}{dt}\Big|_{S'_{\vec{k}}} \frac{\delta H'}{\delta S'_{\vec{k}}} \tag{2.24}$$

From equation (2.22) and (2.23),

$$\frac{dS'_{\vec{k}'}}{dt}\Big|_{S} = (-\frac{1}{2} d - \vec{k}'.\nabla_{\vec{k}'}) S'_{\vec{k}'} \tag{2.25}$$

and

$$\frac{\delta H'}{\delta S_{\vec{k}}} = \exp\{\frac{1}{2} dt\} \frac{\delta H'}{\delta S'_{\vec{k}'}} \tag{2.26}$$

Also,

$$\frac{\partial \alpha'_L(ke^t)}{\partial t} = b + \beta'(k') \tag{2.27}$$

where

$$\beta'(k') = k' \frac{d}{dk'} \beta(k') \tag{2.28}$$

Combining equations (2.24)-(2.28) with equation (2.21), we obtain

$$\begin{aligned} \frac{\partial H'[S']}{\partial t} = \int_{\vec{k}'} & \frac{d}{2} S'_{\vec{k}'} + \vec{k}'.\nabla_{\vec{k}'} S'_{\vec{k}'} \frac{\delta H'}{\delta S'_{\vec{k}'}} \\ & + \int_{\vec{k}'} [b + \beta(k')] \left[\frac{\delta H'}{\delta S'_{\vec{k}'}} \frac{\delta H'}{\delta S'_{-\vec{k}'}} + \frac{\delta^2 H'}{\delta S'_{\vec{k}'} \delta S'_{\vec{k}'}} \right. \\ & \left. + S_{\vec{k}} \frac{\delta H'}{\delta S'_{\vec{k}'}}\right] \end{aligned} \tag{2.29}$$

Equation (2.29) is a slightly general form of Wilson's incomplete-integration renormalization group equation. Wilson chooses $\beta'(k') = 2k'^2$.

2.3 The General Gaussian Renormalization Group Transformation

Wilson's incomplete-integration renormalization group equation can be generalized further. We write the generalized renormalization group equation as:

$$\frac{\partial H'}{\partial t} = \int_{\vec{k}'} \frac{d}{2} S'_{\vec{k}'} + \vec{k}'.\nabla_{\vec{k}'} S'_{\vec{k}} + [b + \beta(k')] S'_{\vec{k}'} \frac{\delta H'}{\delta S'_{\vec{k}'}}$$

$$+ \int_{\vec{k}'} [b + \beta(k')] \gamma(k') \frac{\delta H'}{\delta S'_{\vec{k}'}} \frac{\delta H'}{\delta S'_{-\vec{k}'}} + \frac{\delta^2 H'}{\delta S'_{\vec{k}'} \delta S'_{-\vec{k}'}} \tag{2.30}$$

The function $\gamma(k)$ in equation (2.30) is an arbitrary, analytic, increasing function of k^2. If we set $\gamma(k)$ identically equal to unity, equation (2.30) reduces to the equation (2.29).

A renormalization group transformation can be defined by a differential equation, as in equation (2.30), or alternatively, by functional integral.

$$\exp H'[S'] = \int_{S^o} G(S',S;t) \exp H[S] \tag{2.31}$$

where $G(S', S; t)$ is the Green's function for the differential renormalization group operator. The Green's function for the differential equation (2.30), which we shall construct in the following, is the most general Gaussian Green's function, characterized by two independent parameter-functions, which does not mix spin components of different wavevectors. We, therefore, call the renormalization group equation (2.30), the general Gaussian renormalization group equation.

In order to find the Green's function for the general renormalization group equation, it is convenient to first reduce the equation a differential equation for the functional $e^{H'[S']}$ of the variables k and $S'_{\vec{k}}$ defined in equations (2.22) and (2.23). This is done by retracing the steps made earlier in going from the equation (2.20) to (2.29). It is easy to see that the equation for the functional $e^{H'[S']}$ is

$$\frac{\partial}{\partial t} \exp H'[S']$$

$$= \int_{\vec{k}} [b + \beta(ke^t)] \frac{\partial}{\partial S'_{\vec{k}}} [\gamma(ke^t) \frac{\partial}{\partial S'_{\vec{k}}} + S'_{\vec{k}}] \exp H'[S'] \tag{2.32}$$

Equation (2.32) can be viewed as a functional analog of the Fokker-Planck equation[15]. We know that the Green's function for the Fokker-Planck equation is Gaussian and, therefore, assume that the Green's function for the equation (2.32) is also a Gaussian function. If we replace the integral over $\vec{k}$ on the right-hand-side of equation (2.32) by a discrete sum over a grid we may write

$$\exp H'[S'] = \int_{S^o} G(S',S;t) \text{ ext } H[S] \tag{2.33}$$

with

$$G(S',S;t) = \prod_{\vec{k}} G_{\vec{k}} (S',S;t) \tag{2.34}$$

and

$$G_{\vec{k}} (S',S;t) = \frac{1}{\sqrt{2\pi B(k,t)}} \times \exp \{-\frac{1}{2} \frac{|S'_{\vec{k}} - A(k,t) S_{\vec{k}}|^2}{B(k,t)} \} \tag{2.35}$$

where the functions A(k,t) and B(k,t) are to be determined by the requirement that $G_{\vec{k}}(S',S,t)$ be a solution of the equation

$$\frac{\partial}{\partial t} G_{\vec{k}}(S',S;t) = [b + \beta(ke^t)] \frac{\partial}{\partial S'_{\vec{}}} (\gamma(ke^t) \frac{\partial}{\partial S'_{\vec{k}}}) G_{\vec{k}}(S',S,t) \tag{2.36}$$

and satisfy the initial condition

$$G_{\vec{k}} (S',S;t) = \delta(S'-S) \tag{2.37}$$

We substitute $G_{\vec{k}}(S',S,t)$ from the equation (2.35) in the equation (2.36) and equate the coefficients of equal powers of the spin $S'_{\vec{k}}$ on the two sides of the equation. The equations obtained from equating the linear and the quadratic terms in the spin respectively are

$$\frac{\partial}{\partial t} A(k,t) = - [b + \beta(ke^t)] A(k,t) \tag{2.38}$$

$$\frac{\partial}{\partial t} B(k,t) = 2 [b + \beta(ke^t) \gamma(ke^t) - 2 [b + \beta(ke^t)] B(k,t) \tag{2.39}$$

The equation obtained from equating the constant terms is identical with the equation (2.39).

The solutions of the equations (2.38) and (2.39) are:

$$A(k,t) = C \exp \{-\int_0^t [b + \beta(ke^{t'})] dt'\} \qquad (2.40)$$

where C is an arbitrary constant, and

$$B(k,t) = 2 \int_0^t [b + \beta(ke^{t'})] \gamma(ke^{t'})$$

$$\times \exp \{-2 \int_{t'}^t [b + \beta(ke^{t''})] dt''\} dt' \qquad (2.41)$$

The Green's function $G_{\vec{k}}(S',S,t)$ is now determined. In constructing the Green's function for the equation (2.32) we shall replace the discrete sum over $\vec{k}$ by an integral and ignore the normalization factor $[2\pi B(k,t)]^{-1}$ since it does not affect the calculation of thermodynamic averages. We shall also ignore the constant C in equation (2.40) since the normalization of the variable of integration $S_{\vec{k}}$ in equation (2.33) is unimportant. With this in mind, the integral form for the differential equation (2.36) can be written as

$$\exp H'[S']$$

$$= \int_S \exp \{-\frac{1}{2} \int \frac{|S'_{\vec{k}} - A(k,t)S_{\vec{k}}|^2}{B(k,t)}\} \exp H[S] \qquad (2.42)$$

The integral form for the General Gaussian renormalization group is obtained from equations (2.42), (2.40), and (2.41) by making the scale changes given in equations (2.22) and (2.23). We get

$$\exp H'[S']$$

$$= \int_S \exp \{-\frac{1}{2} \int_{\vec{k}'} \frac{|S'_{\vec{k}'} - \exp\{-\frac{1}{2} dt\} A(k',t) S_{\vec{k}'e^t}|^2}{B(k',t)}\} \exp H[S] \qquad (2.43)$$

where

$$A(k',t) = \exp \{-bt - \int_{k'e^{-t}}^{k'} k''^{-1} \beta(k'') dk''\} \qquad (2.44)$$

$$B(k',t) = 2 \int_{k'e^{-t}}^{k'} [b + \beta(k'')] k''^{-1} \gamma(k'')$$

$$\times \exp \{-2 \int_{k''}^{k'} [b + \beta(k''')] k'''^{-1} dk'''\} dk'' \qquad (2.45)$$

For the choice $\beta(k') = 0$ for $k' \leq 1$, $\beta(k') = \infty$ for $k' > 1$. $\gamma(k') = C$, constant for $k' < 1$, $\gamma(k') = \infty$ for $k' > 1$, equations (2.43), (2.44) and (2.45) can be written as

$$\exp H'[S']$$

$$= \int_S \exp \{-\tfrac{1}{2} C \int_{\vec{k}'} \frac{|S_{\vec{k}'} - \exp\{-\tfrac{1}{2} dt - bt\} S_{\vec{k}'e^{-t}}|^2}{1 - \exp\{-2bt\}}\}$$

$$\times \exp H[S] \tag{2.46}$$

Equation (2.46) is equivalent to the linear renormalization group transformation of Bell and Wilson.[11] When the constant C, in equation (2.46), tends to zero, it becomes equivalent to the sharp cut-off renormalization group of Wilson.[9] An exact differential form for the sharp cutoff renormalization group has been derived by Wegner and Houghton[16]

GAUSSIAN HAMILTONIANS

3.1 The Gaussian Fixed-Point Hamiltonian

The class of renormalization groups which we are considering have the property that a Gaussian Hamiltonian

$$H_o[S^o] = -\frac{1}{2} \int_{\vec{k}} u_2^o(k)\, S^o_{\vec{k}}\, S^o_{-\vec{k}} \tag{3.1}$$

is transformed into a Gaussian Hamiltonian by the renormalization group transformation. For Gaussian Hamiltonians the functional integral over spin in equation (2.43) can be carried out exactly. We change $\vec{k}$ in equation (3.1) to $\vec{k}'$, where $\vec{k}'$ is given by equation (2.22), and also divide and multiply the integrand by $A^2_{\vec{}}(k',t)$. Taking note of the Jacobian of the transformation from $\vec{k}$ to $\vec{k}'$, we obtain

$$H_o[S^o] = -\frac{1}{2} \int_{\vec{k}'} u_2^o(k'e^{-t}) A^{-2}(k';t)$$

$$\times\ |\exp\{-\tfrac{1}{2} dt\}\, A(k',t)\, S^o_{\vec{k}'e^{-t}}|^2 \tag{3.2}$$

We substitute $H_o[S^o]$ from equation (3.2) in the equation (2.43) and apply the rule for the composition of normal distributions.[17] This results in,

$$\exp H_t[S']$$

$$= \exp\{-\frac{1}{2} \int_{\vec{k}'} u_2(k';t)\, S'_{\vec{k}'} S'_{-\vec{k}'} \tag{3.3}$$

where

$$\frac{1}{u_2(k',t)} = B(k',t) + \frac{A^2(k',t)}{u_2^o(k'e^{-t}} \tag{3.4}$$

At the critical point the spin-spin correlation function becomes infinite. The spin-spin correlation function for the Hamiltonian $H_o[S^o]$ is $[u_2^o(k)]^{-1}$ and, therefore, we must have $u_2^o(0) = 0$. We assume that $u_2^o(k)$ possesses the following expansion in k^2, for small k^2,

$$u_2^o(k) = a\ k^2 + O(k^4) \tag{3.5}$$

We substitute the above expansion for $u_2^o(k)$ in equation (3.4) and retain only terms $O(k^2)$. Thus,

$$\frac{1}{u_2(k',t)} = 2 \int_{k'e^{-t}}^{k'} [b + \beta(k'')]\ k''^{-1}\ \gamma(k'') \times \exp\{-2 \int_{k''}^{k'} [b + \beta(k''')]\ k'''^{-1}\ dk'''\} + (a\ k'^2 e^{-2t})^{-1}\ \exp\{-2bt-2 \int_{k'e^{-t}}^{k'} \beta(k'')k''^{-1}dk''\} \tag{3.6}$$

$u_2(k',t)$ approaches a non-trivial fixed point in the limit $t \to \infty$ if we choose

$$b = 1 \tag{3.7}$$

The fixed point is given by,

$$\frac{1}{u_2^*(k')} = \frac{1}{k'^2}\ [2 \int_o^{k'} [1 + \beta(k'')]\ \gamma(k'')k'' \times \exp\{-2 \int_o^{k'} \beta(k''')\ k'''^{-1}\ dk'''\}dk'' + \frac{1}{a} \exp\{-2 \int_o^{k'} \beta(k'')\ k'^{-1}\ dk''\}] \tag{3.8}$$

Equation (3.8) represents a line of fixed points parameterized by the coefficient of k^2 in $u_2^o(k)$. In the renormalization group theory, the critical behavior of the initial Hamiltonian is deduced from the similar behavior of the fixed point Hamiltonian. If b is chosen larger than unity, the limit of H_t would be a Hamiltonian describing a completely non-interacting system of spins and its spin-spin correlation functions would be trivial constants. If b were chosen smaller than unity the limit of H_t would vanish and its spin-spin correlation functions diverge. In neither case would the fixed point Hamiltonian be of any use for obtaining critical exponents. The line of fixed points corresponding to b = 1 are the only useful fixed points of constructing a theory of critical phenomenon[9,11]

3.2 Linear Response of Perturbations to the Critical Hamiltonian, Under Iterated Renormalization Group Transformation

We now discuss the stability of a fixed point Hamiltonian with respect to small perturbations in the initial system.[9] The following discussion is quite general and applies to the Gaussian as well as the more general Hamiltonians which we shall consider in section 4.

Let H^c_o be the critical Hamiltonian of some system which is transformed to the Hamiltonian H^c_t by the renormalization group operation defined by the equation (2.30).

As the renormalization group parameter t tends to infinity, the critical trajectory H^c_t approaches a fixed point Hamiltonian H^*. If a small perturbation δH_o is added to the initial Hamiltonian H^c_o, the trajectory of the renormalized perturbed Hamiltonian H_t would start out with small departure from the critical trajectory H^c_t, that is, for small t, we can write

$$H_t = H^c_t + \delta H_t \tag{3.9}$$

We assume that for large t, when H^c_t approaches H^*, there is a range of t with δH_t still small.[9] We write,

$$H_t = H^* + \delta H_t \tag{3.10}$$

The perturbation δH_t, in equation (3.10), satisfies the equation

$$\frac{\partial}{\partial t} \delta H_t = L\, \delta H_t + O(\delta H_t^2) \tag{3.11}$$

where L is a linear operator and $O(\delta H_t^2)$ refers to terms of second order in the perturbation. The explicit form for the operator L, which depends upon the fixed point Hamiltonian H^*, is obtained by substituting equation (3.10) in the renormalization group equation (2.30). The operator L possesses a set of eigenfunctions $\{O^1_m[S]\}$, with eigenvalues $\{d^1_m\}$, satisfying the equation

$$-d^i_m\, O^i_m[S] = L\, O^i_m[S] \tag{3.12}$$

The set of functionals $\{O^i_m[S]\}$ defines a complete basis set of local operators and the eigenvalues $\{d^i_m\}$ are the anomalous dimensions, or the critical scaling indices, of the local operators.

Each individual functional $O^i_m[S]$, in equation (3.12), is localized about the origin in the coordinate space. One can define a corresponding functional $O^i_m[\vec{x};S]$ centered about an arbitrary point $\vec{x}$ in the coordinate space by the relation

$$O^i_m[x;S] = O^i_m[S'] \tag{3.13}$$

where

$$S'_{\vec{k}} = \exp\{ik.x\}\, S_{\vec{k}} \tag{3.14}$$

If $O^i_m\, S$ is an eigenfunction of the linearlized renormalization group operator L, with the eigenvalue d^i_m, then the translated function $O_m[x\, e^{-t};\, S]$ is also an eigenfunction of L with the eigenvalue d^i_m.

$$-d^i_m\, O^i_m[x\, e^{-t};S] = L\, O^i_m[x\, e^{-t};S] \tag{3.15}$$

We now return to the question of the stability of a fixed point Hamiltonian, when a small perturbation δH_o is added to the initial critical Hamiltonian H^c_o. We write,

$$\delta H_o = \int_{\vec{x}} g(\vec{x})\, O(\vec{x};S) \tag{3.16}$$

where $g(\vec{x})$ is an arbitrary small function and $O(\vec{x};S)$ an arbitrary translated interaction density. We express $O(\vec{x};S)$ as a linear combination of the eigenoperators $O^i_m[\vec{x};\, S]$.

$$O(\vec{x};\, S) = \sum_i C_i\, O^i_m[\vec{x};\, S] \tag{3.17}$$

The solution of equation (3.11) corresponding to the initial condition specified by equations (3.16) and (3.17) is

$$\delta H_t = \sum_i C_i\, \exp\{-d^i_m t\} \int_{\vec{x}} g(\vec{x})\, O^i_m(\vec{x}\, e^{-t};S) \tag{3.18}$$

For translationally invariant perturbations, we must have $g(\vec{x}) = g$, independent of $\vec{x}$. In this case, equation (3.18) simplifies to,

$$\delta H_t = g \sum_i C_i \exp\{(d-d_m^i)t\} \int_{\vec{x}} O_m^i(\vec{x};S) \tag{3.19}$$

It is seen from equation (3.19) that the terms with $d_m > d$ decrease with t while the terms with $d_m < d$ increase with t. The operators $O_m^i[\vec{x}; S]$ with the eingenvalue $d_m^i > d$ are termed as irrelevant operators. The fixed point Hamiltonian and therefore the critical behavior of the Hamiltonian H_o^c remains unchanged under perturbations proportional to the irrelevant operators. The operators with $d_m^i < d$ are called relevant. A perturbation proportional to a relevant operator is associated with changing the temperature of the initial system. When the temperature of the initial system is moved slightly above the critical temperature, the renormalized Hamiltonian develops a term which increases in each iteration of the renormalization group transformation and carries the Hamiltonian farther and farther away from the fixed point. The operators with $d_m^i = d$ are called marginal. For the general Gaussian renormalization group, the marginal operators are associated with changing the renormalization of the spin variables of the original Hamiltonian which may also be parameterized by the variable a which characterizes the line of fixed points for Gaussian Hamiltonians. Perturbations proportional to a marginal operator move the fixed point to a nearby fixed point on the line of the Gaussian fixed points.

3.3 Perturbations to the Gaussian Critical Hamiltonian

The linear renormalization group operator corresponding to the Gaussian fixed point has two relevant eigenoperators which are respectively linear and quadratic in the spin variables. One may generally identify these eigenoperators as the order parameter, or magnetization, and the energy density. We first consider a perturbation linear in spin

$$H_o[S^o] = -\frac{1}{2}\int_{\vec{k}} [u_2^o(k)\, S^o_{\vec{k}}\, S^o_{-\vec{k}} + \varepsilon\, u_1(k)\, S^o_{\vec{k}}] \tag{3.20}$$

where ε is an arbitrary small quantity and $u_1(k)$ an analytic function of k^2. We write

$$u_1(k) = (k^2)^n, \tag{3.21}$$

where n is either zero or a positive integer.

To first order in ε, we can rewrite equation (3.20) as

$$H_o[S^o] = -\tfrac{1}{2}\int_{\vec{k}} \exp\{-\tfrac{1}{2}dt\}\, u_2^o(k'e^{-t})\, S^{o}_{\vec{k}'e^{-t}}\, S^{o}_{-\vec{k}'e^{-t}} \tag{3.22}$$

where

$$S^{o}_{\vec{k}'e^{-t}} = S^{o}_{\vec{k}'e^{-t}} + \varepsilon\, \frac{u_1(k'e^{-t})}{2\, u_2^o(k'e^{-t})} \tag{3.23}$$

The factor $e^{-\frac{1}{2}dt}$, in equation (3.22), comes from the Jacobian of the transformation from the variables $\vec{k}$ to $\vec{k}'$, where $\vec{k}'$ is given by equation (2.22). The purpose behind introducing the variable S is to cast the Hamiltonian in equation (3.20) in a Gaussian form for which we have already determined the fixed point. We define new renormalized spin variables $S_{\vec{k}'}$ by the relation

$$S_{\vec{k}'} = S_{\vec{k}} + \varepsilon\, \frac{\exp\{-\tfrac{1}{2}dt\}\, A(k',t)\, u_1(k'e^{-t})}{2\, u_2^o(k'e^{-t})} \tag{3.24}$$

It is easy to see that

$$S_{\vec{k}'} - \exp\{-\tfrac{1}{2}dt\}\, A(k',t)\, S^{o}_{\vec{k}'e^{-t}}$$

$$= S_{\vec{k}'} - \exp\{-\tfrac{1}{2}dt\}\, A(k',t)\, S^{o}_{\vec{k}'e^{-t}} \tag{3.25}$$

The quantity on the right hand side of equation (3.25) determines the renormalization group transformation kernel in equation (2.43), apart from the factor B(k',t). We see therefore that the kernel of the transformation from the variables S^o to S^o is the same as the kernel of the transformation from the variables S^o to S. Also, the perturbed Hamiltonian has a Gaussian form in the variables S^o similar to the Gaussian form of the old Hamiltonian in the variables S^o. Thus the problem of renormalizing the perturbed Hamiltonian, with perturbation linear in spin, is, to first order in ther perturbation, equivalent to the earlier solved problem of renormalization of a Gaussian Hamiltonian if we replace S^o by S^o and S by S. We obtain,

$$H_t[S] = -\tfrac{1}{2}\int_{\vec{k}'} u_2(k',t)\, S_{\vec{k}'}\, S_{-\vec{k}'} \tag{3.26}$$

where $u_2(k',t)$ is given by equation (3.4). By substituting for $S_{\vec{k}'}$ and $u_1(k'e^{-t})$ from equations (3.24) and (3.21) respectively, we obtain in the limit t tends to infinity and to first order in ε:

$$H_t\ S = H^* - \varepsilon \exp\{-(1+2n)t\} \int_{\vec{k}'} f(k')\ (k')^{2n}\ S_{\vec{k}'} \tag{3.27}$$

where

$$f(k') = \frac{u_2^*(k')}{a\ k'^2} \exp\{-\int_o^{k'} \beta(k'')\ k''^{-1}\ dk''\} \tag{3.28}$$

By comparing equation (3.27) with the equation (3.18), we can identify

$$0_m^{1,n}\ [S] = (k')^{2n}\ f(k')\ S_{\vec{k}'} \tag{3.29}$$

as an eigenoperator of the renormalization group operator linearized about the Gaussian fixed point, with the eigenvalue

$$d_m^{1,n} = (1+2n) \tag{3.30}$$

The superscript 1 on d_m^1 indicates that the eigenoperator corresponding to the eigenvalue d_m^1 is linear in spin. We shall limit ourselves to translationally invariant perturbations, which requires that we must have $n = 0$, and $S_{\vec{k}'} = S_{\vec{k}' = 0}$. The eigenoperator $0_m^1[S]$, with $n = 0$, is a relevant eigenoperator since the eigenvalue associated with it

$$d_m^1 = 1 \tag{3.31}$$

is less than the dimensionality of the system d.

In order to consider perturbations quadratic in spin, we add a small term $\delta u_2^o(k)$ to $u_2^o(k)$ in equation (3.1).

$$u_2^o(k) \rightarrow u_2^o(k) + \delta u_2^o(k) \tag{3.32}$$

$$\delta u_2^o(k) = \varepsilon\ (k^2)^n \tag{3.33}$$

where ε is an arbitrary small number and n any positive integer.

Substitution of equations (3.32) and (3.33) into equation (3.4) results, after some simplifications and taking the limit when t is large, in following

$$H_t \rightarrow H^* - \tfrac{1}{2}\ \varepsilon \exp\{2(1-n)t\} \int_{\vec{k}'} f^2(k')(k')^{2n}\ S_{\vec{k}'}\ S_{-\vec{k}'} \tag{3.34}$$

where $f(k')$ is given by equation (3.28). The quadratic perturbation we consider above is translationally invariant. By comparing equation (3.34) with equation (3.19), we can identify the eigenvalue $d_m^{2,n}$ as

$$d - d_m^{2,n} = 2(1-n)$$

or,

$$d_m^{2,n} = \alpha - 2(1-n) \tag{3.35}$$

The superscript 2 on d_m^2 indicates that the eigenvalue d_m^2 belongs to the eigenvector quadratic in spin. The eigenvector is

$$0_m^{2,n}[S] = f^2(k') \, (k')^{2n} \, S_{\vec{k}'} \, S_{-\vec{k}'} \tag{3.36}$$

It is seen from equation (3.34) that there is one relevant eigenoperator quadratic in spin:

$$0_m^2[S] = f^2(k') \, S_{\vec{k}'} \, S_{-\vec{k}'} \tag{3.37}$$

with the eigenvalue

$$d_m^2 = 2 \tag{3.38}$$

There is one marginal eigenoperator $0_m^{2,n}[S]$ with $n = 1$. All other eigenoperators $0_m^{2,n}$, with $n > 1$, are irrelevant.

3.4 Critical Exponents for Gaussian Hamiltonians

The eigenvalues d_m^1 and d_m^2 of the two relevant eigenoperators respectively determine the set of the critical exponents α, β, γ, δ, Δ, η and ν. The relationships between the critical exponents and the eigenvalues d_m^1 and d_m^2 are:[10,12,18]

$$\nu \, d_m^1 = \beta \tag{3.39}$$

$$\nu \, (d-2 \, d_m^1) = \gamma \tag{3.40}$$

$$\nu \, (d-d_m^1) = \Delta \tag{3.41}$$

$$\nu \, (d-2d_m^2) = \alpha \tag{3.42}$$

$$\nu\ (d-d_m^2) = 1 \tag{3.43}$$

$$(d-2\ d_m^1) = 2-\eta \tag{3.44}$$

$$\frac{(d-d_m^1)}{d_m^1} = \delta \tag{3.45}$$

If we substitute the values of d_m^1 and d_m^2 given by equations (3.31) and (3.38) in the equations (3.39)-(3.45), and set $d = 4$, we obtain;

$$\alpha = 0 \tag{3.46}$$

$$\beta = \frac{1}{2} \tag{3.47}$$

$$\gamma = 1 \tag{3.48}$$

$$\delta = 3 \tag{3.49}$$

$$\Delta = \frac{3}{2} \tag{3.50}$$

$$\eta = 0 \tag{3.51}$$

$$\nu = \frac{1}{2} \tag{3.52}$$

The above predictions of the renormalization group theory for the critical exponents of Gaussian Hamiltonians are identical with the predictions of the Landau[19] and the Ornstein-Zernike[20] theories of phase-transition.

GENERAL HAMILTONIANS

4.1 Renormalization Group Equations for the n-Spin Interaction Functions $u_n(\vec{k},\vec{k}_1,\ldots,\vec{k}_{n-1})$

A general Hamiltonian functional of the fields $S_{\vec{k}}$, which is symmetric under the transformation $S_{\vec{k}} \to -S_{\vec{k}}$, is characterized by an infinite set of functions $u_2(\vec{k})$, $u_4(\vec{k},\vec{k}_1,\vec{k}_2,\vec{k}_3)$, $u_6(\vec{k},\vec{k}_1,\vec{k}_2,\vec{k}_3,\vec{k}_4,$

$\vec{k}_5$),..., etc, as shown in equation (2.9).

We substitute equation (2.9) in the general Gaussian renormalization group equation (2.30) and obtain the following differential equations for the u-functions: (the primes on S, $\vec{k}$ and $\beta(k)$ are omitted for convenience).

$$\frac{\partial u_2}{\partial t}(\vec{k},t) = \{-\vec{k}.\nabla_{\vec{k}} + 2[b + \beta(k)]$$

$$\times\ [1-\gamma(k)u_2(\vec{k},t)]\ u_2(\vec{k},t)$$

$$+ \int_{\vec{k}_1} [b + \beta(k_1)]\gamma(k_1)\ u_4(\vec{k},-\vec{k},\vec{k}_1,-\vec{k}_1,t) \tag{4.1}$$

$$\frac{\partial u_4}{\partial t}(\vec{k},\vec{k}_1,\vec{k}_2,\vec{k}_3,t) = d + \sum_i A(k_i,t)]$$

$$\times\ u_4(\vec{k},\vec{k}_1,\vec{k}_2,\vec{k}_3,t)$$

$$+ \int_{\vec{k}_4} [1+\beta(k_4)]\gamma(k_4)\ u_6(\vec{k},\vec{k}_1,\vec{k}_2,\ \vec{k}_3,\vec{k}_4,-\vec{k}_4,t) \tag{4.2}$$

$$\frac{\partial u_6}{\partial t}(\vec{k},\vec{k}_1,\vec{k}_2,\vec{k}_3,\vec{k}_4,\vec{k}_5,t) = [d + \sum_i A(\vec{k}_i,t)\ \ u_6(\vec{k},\ldots,\vec{k}_5,t)$$

$$- 2\sum_p [b + \beta(\vec{k} + \vec{k}_1 + \vec{k}_2)]\gamma(\vec{k} + \vec{k}_1 + \vec{k}_2)$$

$$\times\ u_4(\vec{k},\vec{k}_1,\vec{k}_2,\vec{k}_3 + \vec{k}_4 + \vec{k}_5,t)\ u_4(\vec{k}_3,\vec{k}_4,\vec{k}_5,\vec{k} + \vec{k}_1 + \vec{k}_2,t)$$

$$+ \int_{\vec{k}_6} [b + \beta(k_6)]\gamma(k_6)\ u_8(\vec{k},\ldots,\vec{k}_5,\vec{k}_6,-\vec{k}_6,t) \tag{4.3}$$

In general, for n = 4, 5, 6, ...,

$$\frac{\partial u_{2n}}{\partial t}(\vec{k},\ldots,\ \vec{k}_{2n-1},\ t) = [d + \sum_i A(\vec{k}_i,t)]\ u_{2n}(\vec{k},\ldots,\ \vec{k}_{2n-1},t)$$

$$- 2 \sum_{p}^{<\frac{1}{2}(n+1)} \sum_{p=2} [b + \beta(\vec{k} + \ldots + k_{2\ p-2})]\gamma(\vec{k} + \ldots + \vec{k}_{2\ p-2})$$

$$\times\ u_{2p}(\vec{k}, \ldots, \vec{k}_{2\ p-2}, \vec{k}_{2\ p-1} + \ldots + \vec{k}_{2n-1}, t)$$

$$\times\ u_{2n+2-2p}(\vec{k}_{2p-1}, \ldots, \vec{k}_{2n-1}, \vec{k} + \ldots + \vec{k}_{2p-2}, t)$$

$$+ \int_{\vec{k}_{2n}} [b + \beta(k_{2n})]\gamma(k_{2n})\ u_{2n+2}(k, \ldots, k_{2n-1}, k_{2n}, -k_{2n}, t) \qquad (4.4)$$

where

$$A(\vec{k}_{i,t}) = - k_i . \nabla_{\vec{k}_i} - \frac{d}{2} + [b + \beta(k_i)]$$

$$\times\ \{1 - 2\gamma(k_i)\ u_2(k_i, t)\} \qquad (4.5)$$

$\sum_{i} A(\vec{k}_i, t)$ denotes the sum of operators $A(\vec{k}_i, t)$ over all wavevectors $\vec{k}_i$, for example, $\vec{k}_i = \vec{k}, \vec{k}_1, \ldots, \vec{k}_{2n-1}$ in equation (4.4). $\sum_{p}$ denotes the sum over inequivalent permutations of $(\vec{k}, \vec{k}_1, \ldots, \vec{k}_{2n-1})$ in the $u_{2p} \times u_{2n+2-p}$ term.

The ordinary partial differential equations (4.1)-(4.4) are completely equivalent to the functional differential equation (2.30) for the general Hamiltonian given by the equation (2.9).

4.2 The Epsilon Expansion

We are unable to solve the infinite hierarchy of equations (4.1)-(4.4), for arbitrary interactions $u_2(\vec{k})$, $u_4(\vec{k}, \vec{k}_1, \vec{k}_2, \vec{k}_3)$, ... etc. We have, however, seen that if the non-Gaussian interactions u_4, u_6,... etc., are initially zero, the equations (4.1)-(4.4) possess a solution which approaches a fixed point, namely, the Gaussian fixed point, as t tends to infinity. We assume that when the non-Gaussian interactions are small, the equations (4.1)-(4.4) possess a general fixed point in the neighborhood of the Gaussian fixed point and that the general fixed point approaches the Gaussian fixed point as the non-Gaussian interactions u_4, u_6,... etc., approach zero.

We shall denote the general fixed point interaction functions by $u^*_{2n}(\vec{k}, \vec{k}_1, \ldots, \vec{k}_{2n-1})$. The functions $u^*_{2n}(\vec{k}, \vec{k}_1, \ldots, \vec{k}_{2n-1})$

must be independent of t:

$$\frac{\partial}{\partial t} u^*_{2n}(k, \ldots, k_{2n-1}) = 0 \tag{4.6}$$

We assume that the non-Gaussian interactions u^*_4, u^*_6,... etc., are at most of the order of ε, where ε is an arbitrary small quantity, and write

$$b = 1 + b_1\,\varepsilon + b_2\,\varepsilon^2 + \ldots \tag{4.7}$$

$$u^*_2(k) = u^*_{20}(k) + u^*_{21}(k)\,\varepsilon + u^*_{22}(k)\,\varepsilon^2 + \ldots \tag{4.8}$$

$$u^*_4(\{\vec{k}_i\}) = u^*_{41}(\{\vec{k}_i\})\,\varepsilon + u^*_{42}(\{\vec{k}_i\})\,\varepsilon^2 + \ldots \tag{4.9}$$

$$u^*_6(\{\vec{k}_i\}) = u^*_{61}(\{\vec{k}_i\})\,\varepsilon + u^*_{62}(\{\vec{k}_i\})\,\varepsilon^2 + \ldots \tag{4.10}$$

We have chosen $b = 1$ for $\varepsilon = 0$, in equation (4.7) in view of the equation (3.7) for Gaussian Hamiltonians. $u^*_{20}(k)$, in equation (4.8), denotes the Gaussian fixed point interaction. The expansion in equations (4.7)-(4.10) would not have much practical value unless it helps us to truncate the infinite hierarchy of equations (4.1)-(4.4) to some finite order in ε. We now investigate this possibility. Let us assume that to some order in ε, the u^*_8 term in equation (4.3) can be neglected in comparison with the other terms. Assuming the interaction functions are analytic at $\{\vec{k}_i\} = 0$, we obtain equation (4.3),

$$(6-2d)\, u^*_6(\{0\})$$

$$= - 20\, \gamma(0)\, u^*_4(\{0\}) \times u^*_4(\{0\}) \tag{4.11}$$

We shall see in the following that the functions $[u^*_2(k)-u^*_{20}(k)]$, $u^*_4(\vec{k}, \vec{k}_1, \vec{k}_2, \vec{k}_3)$, $u^*_6(\vec{k}, \vec{k}_1, \vec{k}_2, \vec{k}_3, \vec{k}_4, \vec{k}_5)$ ... etc., are decreasing functions of the wavevectors $\{\vec{k}_i\}$ and, therefore, the order of magnitude of these functions can be estimated by setting all the argument wavevectors equal to zero. Since u^*_4 is at most of the order of ε, u^*_6, according to the equation (4.11), must be at most of the order ε^2. Thus, we must have,

$$u^*_{61}(\{k_i\}) \equiv 0 \tag{4.12}$$

Again, assuming the functions $u^*_2(k)$ and $u^*_4(\{k_i\})$ are analytic at the origin, we can write the following relations from the equations

(4.1) and (4.2)

$$u^*_{21}(0)$$

$$= -\tfrac{1}{2}\int_{\vec{k}_1} [1+\beta(k_1)]\gamma(k_1)\; u^*_{41}(0,0,\vec{k}_1,-\vec{k}_1) \tag{4.13}$$

$$(4-d)\; u^*_{41}(\{0\})$$

$$= \int_{\vec{k}_4} [1+\beta(k_4)]\gamma(k_4)\; u^*_{62}(0,0,0,0,\vec{k}_4,-\vec{k}_4) \tag{4.14}$$

Equation (4.14) determines the nature of the parameter ε. If u^*_6 is at most of the order ε^2 and u^*_4 of the order of ε, then equation (4.14) can only be satisfied if ε is taken to be proportional to (4-d) for small ε. For simplicity, we make the identification

$$\varepsilon = 4-d \tag{4.15}$$

In the following we assume that ε is small and positive, that is to say the systems we shall be analyzing have slightly less than four dimensions. The idea of non-integer dimension $d = 4-\varepsilon$ and an expansion around $d = 4$ serves to construct an analytic continuation of the Gaussian fixed point. The epsilon expansion is a calculational device which has been used elsewhere in statistical mechanics and quantum field theory[21]. The use of such a device does not, of course, imply that non-integer dimensions have any physical significance.

The equations (4.11)-(4.15) were derived on the assumption the u^*_8 term in equation (4.3) can be neglected. An explicit study of the u^*_8 equation reveals that u^*_8 is of the order of ε^3. This justifies our assumption that, to order ε^2, the u^*_8 term in the equation (4.3) can be neglected. The fixed point interaction functions u^*_{20}, u^*_{21}, u^*_{22}, u^*_{41}, u^*_{42} and u^*_{62}, in equations (4.8)-(4.10), satisfy the following equations:

$$\vec{k}.\nabla_{\vec{k}}\, u^*_{20}(\vec{k}) = 2[1+\beta(k)]\,[1-u^*_{20}(\vec{k})\gamma(k)]\; u^*_{20}(\vec{k}) \tag{4.16}$$

$$\vec{k}.\nabla_{\vec{k}}\, u^*_{21}(\vec{k}) = 2[1+P(\vec{k})]\; u^*_{21}(\vec{k}) + 2b_1[1-u^*_{20}(\vec{k})\gamma(k)] \times u^*_{20}(\vec{k}) + \int_{\vec{k}} [1+\beta(k_1)]\gamma(k_1)u^*_{41}(\vec{k},-\vec{k},\vec{k}_1,-\vec{k}_1) \tag{4.17}$$

where

$$P(k) = \beta(k)[1-2\ u^*_{20}(k)\gamma(k)]\ -2\ u^*_{20}(k)\gamma(k) \tag{4.22}$$

The integrals are four-dimensional. $\int^{\varepsilon}_{\vec{k}}$ denotes the term of the order of ε in the expansion of $\int_{\vec{k}}$ around dimensionality four.

4.3 The General Fixed Point Hamiltonian in Four Dimensions

The general fixed point Hamiltonian in four dimensions, that is, for $\varepsilon = 0$, has only one term in it, namely, $u^*_{20}(\vec{k})$. This, of course, must be identical with the $u^*_2(k)$ for the Gaussian fixed point given in equation (3.8). It is instructive, however, to determine $u^*_{20}(k)$ by solving equation (4.16) directly. We assume that $u^*_{20}(\vec{k})$ depends only on $|k|$ and accordingly replace $\vec{k}.\nabla_{\vec{k}}u_{20}(\vec{k})$, in equation (4.16), by $k\frac{d}{dk}u^*_{20}(k)$. We introduce an auxiliary function $\Gamma(k)$ through the relation

$$u^*_{20}(k) = \bar{u}^*_{20}(k)\ \times\ \Gamma(k) \tag{4.23}$$

Substituting (4.23) in (4.16),

$$\begin{aligned}\bar{u}^*_{20}(k)\ k\frac{d}{dk}\Gamma(k) + \Gamma(k)\ k\frac{d}{dk}\bar{u}^*_{20}(k) \\ = 2[1+\beta(k)]\gamma(k)\ \bar{u}^*_{20}(k) \\ - 2[1+\beta(k)]\gamma(k)\ \Gamma^2(k)\ \bar{u}^{*\,2}_{20}(k)\end{aligned} \tag{4.24}$$

We require the auxiliary function $\Gamma(k)$ to satisfy,

$$k\frac{d}{dk}\Gamma(k) = 2[1+\beta(k)]\ \Gamma(k) \tag{4.25}$$

With the condition (4.25), equation (4.24) reduces to,

$$k\frac{d}{dk}\bar{u}^*_{20}(k) = -2[1+\beta(k)]\gamma(k)\ \Gamma(k)\ \bar{u}^{*\,2}_{20}(k) \tag{4.26}$$

Equation (4.25) can be rewritten as

$$\vec{k}\cdot\nabla_{\vec{k}}\, u^*_{22}(\vec{k}) = 2[1+P(k)]\, u^*_{22}(\vec{k}) - 2[1+\beta(k)]\gamma(k)$$

$$\times\, u^*_{20}(\vec{k})\, u^*_{21}(\vec{k}) + 2b_1[1-2u^*_{20}(\vec{k})\gamma(k)]$$

$$\times\, u^*_{21}(\vec{k}) + 2b_2[1-u^*_{20}(\vec{k})\gamma(k)]\, u^*_{20}(\vec{k})$$

$$+ \int_{\vec{k}_1} [1+\beta(k_1)]\gamma(k_1)\, u^*_{42}(\vec{k},-\vec{k},\vec{k}_1,-\vec{k}_1)$$

$$+ \int^{\varepsilon}_{\vec{k}_1} [1+\beta(k_1)]\gamma(k_1)\, u^*_{41}(\vec{k},-\vec{k},\vec{k}_1,-\vec{k}_1) + b_1$$

$$\times \int_{\vec{k}_1} \gamma(k_1)\, u^*_{41}(\vec{k},-\vec{k},\vec{k}_1,-\vec{k}_1) \qquad (4.18)$$

$$\sum_i \vec{k}_i\cdot\nabla_{\vec{k}_i}\, u^*_{41}(\vec{k},\vec{k}_1,\vec{k}_2,\vec{k}_3) = \sum_i P(\vec{k}_1)\, u^*_{41}(\{k_i\}) \qquad (4.19)$$

$$\sum_i \vec{k}_i\cdot\nabla_{\vec{k}_i}\, u^*_{42}(\vec{k},\vec{k}_1,\vec{k}_2,\vec{k}_3) = \sum_i P(\vec{k}_1)\, u^*_{42}(\{\vec{k}_i\}) + u^*_{41}(\{\vec{k}_i\})$$

$$+ \sum_i \{b_1[1-2\, u^*_{20}(\vec{k}_i)\gamma(k_i)] - [1+\beta(k_1)]\gamma(k_1)$$

$$\times\, u^*_{21}(\vec{k}_i)\}\, u^*_{41}(\{\vec{k}_i\})$$

$$+ \int_{\vec{k}_4} [1+\beta(k_4)]\gamma(k_4)$$

$$\times\, u^*_{62}(\vec{k},\vec{k}_1,\vec{k}_2,\vec{k}_3,\vec{k}_4,-\vec{k}_4) \qquad (4.20)$$

$$\sum_i \vec{k}_1\cdot\nabla_{\vec{k}_1}\, u^*_{62}(\vec{k},\vec{k}_1,\vec{k}_2,\vec{k}_3,\vec{k}_4,\vec{k}_5) = -\,2\, u^*_{62}(\{\vec{k}_i\})$$

$$+ \sum_i P(k_i)\, u^*_{62}(\{\vec{k}_i\})$$

$$-\,2 \sum_P \lfloor 1+\beta(\vec{k}+\vec{k}_1+\vec{k}_2)\rfloor\gamma(\vec{k}+\vec{k}_1+\vec{k}_2)$$

$$\times\, u^*_{41}(\vec{k},\vec{k}_1,\vec{k}_2,\vec{k}_3+\vec{k}_4+\vec{k}_5)$$

$$\times\, u^*_{41}(\vec{k}_3,\vec{k}_4,\vec{k}_5,\vec{k}+\vec{k}_1+\vec{k}_2) \qquad (4.21)$$

$$\frac{d\,\bar{u}^*_{20}}{\bar{u}^{*\,2}_{20}} = -\frac{2[1+\beta(k)]\gamma(k)\ \Gamma(k)}{k}\,dk \tag{4.27}$$

Integration of equation (4.27) yields,

$$-\frac{1}{\bar{u}^*_{20}(k)} = -2\int^k \frac{[1+\beta(k')]\gamma(k')\ \Gamma(k')}{k'}\,dk' \tag{4.28}$$

or,

$$\bar{u}^*_{20}(k) = [\frac{1}{a} + \int_o^k \frac{2[1+\beta(k')]\gamma(k')\ \Gamma(k')}{k'} \tag{4.29}$$

where a is an arbitrary constant. Equation (4.25) for $\Gamma(k)$ can be solved as follows; write,

$$\frac{d\ \Gamma(k)}{\Gamma(k)} = 2\ \frac{[1+\beta(k)]}{k}dk \tag{4.30}$$

and integrate equation (4.30) to obtain,

$$\ell n\ \Gamma(k) = \ell n\ k^2 + 2\int^k \frac{\beta(k')}{k'}\,dk' \tag{4.31}$$

or,

$$\Gamma(k) = k^2\ \exp\ \{2\int_o^k \frac{\beta(k')}{k'}\,dk'\} \tag{4.32}$$

We have chosen the coefficient of k^2 in $\Gamma(k)$ equal to unity. This, however, does not result in any loss of generality since the coefficient of k^2 in $u^*_{20}(k)$ is arbitrary. Combining equations (4.29) and (4.32), we obtain

$$\begin{aligned} u^*_{20}(k) = k^2[a^{-1}\ &\exp\{-2\int_o^k k'^{-1}\beta(k')dk'\} \\ &+ 2\int_o^k [1+\beta(k')\gamma(k')k' \\ &\times \exp\{-2\int_{k'}^k k''^{-1}\ \beta(k'')dk''\}dk']^{-1} \end{aligned} \tag{4.33}$$

4.4 The General Fixed Point Hamiltonian in (4-ε) Dimensions, to Order ε.

The general fixed point Hamiltonian to order ε, is specified by the interaction functions u^*_{41} and u^*_{21}.

The function u^*_{41} is determined by the equation (4.19). We shall solve (4.19) by the method of characteristics.[22] In Fourier space, the characteristic of equation (4.19) which passes through the configuration $\{\vec{k}_i\}$ is the one-dimensional set of points $\lambda\{\vec{k}_i\}$ for all positive values of λ. To determine how $u^*_{41}(\{\vec{k}_i\}$ changes on such a characteristic we substitute $\lambda\{\vec{k}_i\}$ for $\{\vec{k}_i\}$ and note that the differential operator $\sum_i \vec{k}_i \cdot \nabla_{\vec{k}_i}$ can be replaced by $\lambda \frac{d}{d\lambda} u^*_{41}(\{\vec{k}_i\})$.
Thus we obtain an ordinary differential equation for $u^*_{41}(\lambda\{\vec{k}_i\})$, $\{\vec{k}_i\}$ fixed,

$$\lambda \frac{d}{d\lambda} u^*_{41}(\lambda\{\vec{k}_i\}) = \sum_i P(\lambda k_i)\; u^*_{41}(\lambda\{k_i\}) \tag{4.34}$$

It is seen from equation (4.34) that the method of characteristics transforms equation (4.19) involving the operator $\sum_i \vec{k}_i \cdot \nabla_{\vec{k}_i}$ into a much simpler one-dimensional equation. The operator $\sum_i \vec{k}_i \cdot \nabla_{\vec{k}_i}$ occurs in all the equations which arise in the course of our analysis and, therefore, we shall use the method of characteristics extensively. Equation (4.34) is a particular form of the general inhomogeneous linear differential equation of the first order:

$$\frac{d}{d\lambda} u^*(\lambda) = y(\lambda)\; u^*(\lambda) + z(\lambda) \tag{4.35}$$

The solution of equation (4.35) is

$$u^*(\lambda) = u^*(\xi)\; \exp\{\int_\xi^\lambda y(\lambda')d\lambda'\} + \int_\xi^\lambda z(\lambda')\; \exp\{\int_{\lambda'}^\lambda y(\lambda'')d\lambda''\}d\lambda' \tag{4.36}$$

where ξ is an arbitrary constant. In using equation (4.36) to obtain the solution for (4.34) we choose ξ = 0;

$$u^*_{41}(\lambda\{\vec{k}_i\}) = A\; \exp\{\sum_i \int_o^\lambda \frac{P(\lambda' k_i)}{\lambda'}\, d\lambda'\} \tag{4.37}$$

where $A = u^*_{41}(\{0\})$. The constant A is not determined by the equation (4.19) or (4.34). We set λ = 1, to obtain

$$u^*_{41}(\{\vec{k}_i\}) = A \exp\{ \sum_i \int_o^1 \frac{P(\lambda' k_i)}{\lambda'} d\lambda\} \tag{4.38}$$

Equation (4.38) can be rewritten as

$$u^*_{41}(\{\vec{k}_i\}) = A \prod_i f(k_i) \tag{4.39}$$

with

$$f(k) = \exp\{ \int_o^k \frac{P(k')}{k'} dk'\} \tag{4.40}$$

Next, we determine $u^*_{21}(k)$; combining equations (4.13), (4.39) and (4.40) gives,

$$u^*_{21}(0) = - \frac{1}{2} A B \tag{4.41}$$

where

$$B = \int_{\vec{k}} \psi(k) \tag{4.42}$$

$$\psi(k) = [1 + \beta(k)]\gamma(k)f^2(k) \tag{4.43}$$

The general solution for $u^*_{21}(k)$ can now be written after the model equation (4.36);

$$\begin{aligned} u^*_{21}(k) = - \frac{1}{2} A B \exp\{2 \int_o^1 \frac{[1+P(\lambda k)]}{\lambda} d\lambda \\ + \int_o^1 \lambda'^{-1} \{2b_1[1-u^*_{20}(\lambda' k)\gamma(\lambda' k)]\, u^*_{20}(\lambda' k) \\ + A\, B\, f^2(\lambda' k)\} \exp\{2\int_{\lambda'}^1 \lambda''^{-1} \\ \times [1+P(\lambda'' k)]d\lambda''\}\, d\lambda' \end{aligned} \tag{4.44}$$

we write $f(\lambda' k)$ as,

$$f(\lambda' k) = \exp\{ \int_o^{\lambda'} \lambda''^{-1} P(\lambda'' k)d\lambda''\} \tag{4.45}$$

and combine it with a part of the exponential term in the integrand. This results in

$$u^*_{21}(k) = -\frac{1}{2} A B \exp\{ 2 \int_o^1 \lambda^{-1}[1+P(\lambda k)]d\lambda\}$$

$$+ A B \exp\{ 2 \int_o^1 \lambda^{-1} P(\lambda k)d\lambda\} \int_o^1 \lambda'^{-1}$$

$$x \exp\{ \int_{\lambda'}^1 \lambda''^{-1} d\lambda''\}d\lambda' + 2b_1 \int_o^1 \lambda'^{-1}$$

$$\{1-u^*_{20}(\lambda' k)\gamma(\lambda' k)\} u^*_{20}(\lambda' k) \exp\{2\int_{\lambda'}^1 \lambda''^{-1}[1+P(\lambda'' k)]d\lambda''\}d\lambda'$$

$$= -\frac{1}{2} A B f^2(k) + 2b_1\{....\} \tag{4.46}$$

where we have used the fact that

$$[-\frac{1}{2} \exp\{2 \int_o^1 \lambda^{-1} d\lambda\} + \int_o^1 \lambda'^{-1} \exp\{2 \int_{\lambda'}^1 \lambda''^{-1} d\lambda''\}d\lambda']$$

$$= \lim_{\delta\to 0} [-\frac{1}{2} \exp\{2 \int_\delta^1 \lambda^{-1} d\lambda\} + \int_\delta^1 \lambda'^{-1} \exp\{\ell n(\lambda')^{-2}\}d\lambda']$$

$$= -\frac{1}{2} \tag{4.47}$$

The function appearing with b_1 in equation (4.46) is not regular at $\vec{k} = 0$ and therefore we must set,

$$b_1 = 0 \tag{4.48}$$

to assure that $u^*_{21}(k)$ is analytic at $k = 0$. The most general analytic solution for the function $u^*_{21}(k)$ is obtained by adding to the right hand side of equation (4.46) an arbitrary multiple of the solution of the homogeneous part of the equation (4.17):

$$u^*_{21}(k) = (-\frac{1}{2} A B + C_1 k^2) f^2(k) \tag{4.49}$$

where C_1 is an arbitrary constant.

4.5 The General Fixed Point Hamiltonian in (4 - ε) Dimensions, to Order ε^2.

To determine the general fixed point Hamiltonian to order ε^2, we have to solve equations (4.18), (4.20), and (4.21) for the interaction function $u^*_{22}(k)$, $u^*_{42}(\vec{k}, \vec{k}_1, \vec{k}_2, \vec{k}_3)$ and $u^*_{62}(\vec{k}, \vec{k}_1, \vec{k}_2, \vec{k}_3,$

$\vec{k}_4$, $\vec{k}_5$). We solve the equation (4.21) first. Assuming that $u^*_{62}(\{\vec{k}_i\})$ is analytic in $\{\vec{k}_i\}$ at $\{\vec{k}_i\} = 0$, equation (4.21) yields,

$$u^*_{62}(\{0\}) = -\ 10\ \gamma(10)\ A^2 \tag{4.50}$$

Equation (4.21) is a linear inhomogeneous differential equation of the first order. Earlier, in equation (4.36), we have given the solution of such an equation with arbitrary coefficient functions. Substituting the appropriate functions for $y(\lambda)$ and $z(\lambda)$ in equation (4.36), and putting the boundary condition (4.50 in, we obtain

$$\begin{aligned} u^*_{62}(\{k_i\}) = &-\ 10\ \gamma(0)\ A^2\ \exp\{ \int_o^1 \lambda^{-1}[-2+\sum_i P(\lambda k_i)] \\ &\times d\lambda\}\ -2A^2 \int_o^1 \lambda'^{-1}\ P_1(\lambda'\{\vec{k}_i\}) \\ &\times \exp\{ \int_\lambda^1 \lambda''^{-1}\ [-2+\sum_i P(\lambda'' k_i)]d\lambda' \end{aligned} \tag{4.51}$$

where $P_1(\lambda'\{k_i\})$ corresponds to the inhomogeneous term in equation (4.21).

$$\begin{aligned} P_1(\lambda'\{k_i\}) = &\sum_P [1+\beta(\lambda'\vec{k}+\lambda'\vec{k}_1+\lambda'\vec{k}_2)]\gamma(\lambda'\vec{k}+\lambda'\vec{k}_1+\lambda'\vec{k}_2) \\ &\times f(\lambda'k)f(\lambda'k_1)\ f(\lambda'k_2)\ f(\lambda'k+\lambda'k_1+\lambda'\vec{k}_1+\lambda'\vec{k}_2) \\ &\times f(\lambda'k_3)f(\lambda'k_2)f(\lambda'k_5)f(\lambda'k_3+\lambda'k_4+\lambda'k_5) \end{aligned} \tag{4.52}$$

Equations (4.51) and (4.52) can be combined to give

$$\begin{aligned} u^*_{62}(\{k_i\}) = -A^2 \prod_i f(k_i)\ &[10\gamma(0)\ \exp\{-2\int_{\lambda'}^1 \lambda^{-1}d\lambda\} \\ &\times 2\int_0^1 \lambda'^{-1}P_2(\lambda'\{\vec{k}_i\})\ \exp\{-2\int_{\lambda'}^1 \lambda''^{-1}d\lambda'\}d\lambda' \end{aligned} \tag{4.53}$$

where,

$$\begin{aligned} P_2(\lambda'\{\vec{k}_i\}) = &\sum_P [1+\beta(\lambda^1\vec{k}+\lambda^1\vec{k}_1+\lambda'\vec{k}_2)] \\ &\times \gamma(\lambda'\vec{k}+\lambda'\vec{k}_1+\lambda'\vec{k}_2)f^2(\lambda'\vec{k}+\lambda'\vec{k}_1+\lambda'\vec{k}_2) \end{aligned} \tag{4.54}$$

or,

$$P_2(\lambda'\{\vec{k}_i\}) = \sum_P \psi(\lambda'\vec{k}+\lambda'\vec{k}_1+\lambda'\vec{k}_2) \tag{4.55}$$

The first term in equation (4.53) contributes zero. Performing the integral over λ'' in the second term, we obtain

$$u^*_{62}(\{\vec{k}_i\}) = -2A^2 \prod_i f(k_i) \sum_P G(\vec{k}+\vec{k}_1+\vec{k}_2) \tag{4.56}$$

where,

$$g(k) = \int_0^1 \lambda\ \psi(\lambda k)\ d\lambda \tag{4.57}$$

We now proceed to determine $u^*_{42}(\{\vec{k}_i\})$. We assume that $u^*_{42}(\{\vec{k}_i\})$ is an analytic function of $\{\vec{k}_i\}$ and for small $\vec{k}_i$, we can write

$$u^*_{42}(\{\vec{k}_i\}) = u^*_{42}(\{0\}) + \text{const.}\ \vec{k}_i.\vec{k}_j + O(k^4) \tag{4.58}$$

We substitute (4.58) in the equation (4.20) and then equate the constant terms on both sides of the equation. This gives

$$A+4A^2B+\int_{\vec{k}_4} [1+\beta(k_4)]\gamma(k_4)\ u^*_{62}(0,0,0,0,\vec{k}_4,-\vec{k}_4) = 0 \tag{4.59}$$

The constant $u^*_{42}(\{0\})$ is not determined by the equation (4.20), just as $u^*_{41}(\{0\})$ was not determined by (4.19). Substituting,

$$u^*_{62}(0,0,0,0,k_4,-k_4) = -\ 2A^2f^2(k_4)\ [2+6g(k_4)] \tag{4.60}$$

in the equation (4.59), we obtain

$$A = [12 \int_{\vec{k}_4} \psi(k_4)g(k_4)]^{-1} \tag{4.61}$$

The solution for u^*_{42} can be obtained in the same manner as u^*_{21} and u^*_{62} were obtained. The algebra is very similar, though lengthier. We obtain

$$u^*_{42}(\{k_i\}) = \prod_{i=0}^{3} f(k_i)[C_2+A^2B \sum_{i=0}^{3} R(k_i) -2AC_1$$
$$\times \sum_{i=0}^{3} k_i^2 g(k_i) - \frac{2}{3} A^2B\{3S(k)+\sum_{i=1}^{3}$$

$$S(k_i) + \sum_{3P} S(\vec{k} + \vec{k}_i + \vec{k}_j) -$$

$$\times \sum_{6P} T(k_i + k_j) - \frac{4}{3} A^2 \qquad T(\vec{k} + \vec{k}_i)] \tag{4.62}$$

where

$$R(k) = \int_0^1 \lambda^{-1} [\psi(\lambda k)-1]d\lambda \tag{4.63}$$

$$S(k) = \int_0^1 \lambda^{-1} [g(\lambda k) - \frac{1}{2}]d\lambda \tag{4.64}$$

$$T(k) = \int_0^1 \lambda^{-1} [G(\lambda k) - G(0)]d\lambda \tag{4.65}$$

$$G(k) = 2 \int_{\vec{k}'} \psi(k')g(\vec{k} + \vec{k}') \tag{4.66}$$

and C_2 is an arbitrary constant.

Finally, we come to the equation (4.18). We assume that $u^*_{22}(k)$ is analytic in k^2 and that for small k, we can write

$$u^*_{22}(k) = u^*_{22}(0) + C_3 k^2 + O(k^4) \tag{4.67}$$

where C_3 is an arbitrary constant. We substitute the expansion (4.67) in the equation (4.18) and equate the constant terms and the coefficients of k^2 on the two sides of that equation. This results in the following equations (we use $b_1 = 0$, equation (4.48)).

$$2 u^*_{22}(0) = 2 \gamma(0) u^*_{21}{}^2(0) - \int_{\vec{k}_1} [1+\beta(k_1)]\gamma(k_1) u^*_{42}(0,0,\vec{k}_2,-\vec{k}_2)$$

$$= \int_{\vec{k}_1}^{\varepsilon} [1+ (k_1)]\beta(k_1)\gamma u^*_{41}(0,0,\vec{k}_1,-\vec{k}_1) \tag{4.68}$$

$$- 2b_2 a = 2 u^*_{22}(0)[\beta_2-2a\gamma(0)] - 2[2\gamma(0) u^*_{21}(0)C_4$$

$$+ \{\beta_2\gamma(0) + \gamma_2\} u^*_{21}{}^2(0)] - \frac{1}{2} \int_{\vec{k}_1} [1+\beta(k_1)]$$

$$\times \gamma(k_1)\, \nabla_{\vec{k}} . \nabla_{\vec{k}}\, u^*_{42}(\vec{k},-\vec{k},\vec{k}_1,-\vec{k}_1)|_{\vec{k}=0}$$

$$+ \int_{\vec{k}_1} [1+\beta(k_1)]\gamma(k_1)\, \nabla_{\vec{k}} \cdot \nabla_{\vec{k}}\, u^*_{41}(\vec{k},-\vec{k},\vec{k}_1,-\vec{k}_1)|_{\vec{k}=0} \tag{4.69}$$

where,

$$\gamma_2 = \frac{1}{2}\frac{d^2}{dk^2}\gamma(k)|_{k=0} \tag{4.70}$$

$$\beta_2 = \frac{1}{2}\frac{d^2}{dk^2}\beta(k)|_{k=0} \tag{4.71}$$

and C_4 is the coefficient of k^2 in the expansion of $u^*_{21}(k)$ for small k. We substitute the value of $u^*_{22}(0)$ given by the equation (4.68) in the equation (4.69) and, after some simplification, obtain

$$-2b_2 = -\frac{A^2}{a}\int_{\vec{k}_2}\psi(k_2)\int_{\vec{k}_4}\psi(k_4)\int_0^1 \lambda\,\frac{1}{F}\frac{d}{dF}\psi(F)d\lambda \tag{4.72}$$

where

$$F = |\lambda\vec{k}_2 + \vec{k}_4| \tag{4.73}$$

The simplification we alluded to above proceeds as follows. We write,

$$u^*_{42}(\vec{k},-\vec{k},\vec{k}_1,-\vec{k}_1) = f^2(k)f^2(k_1)\bar{u}^*_{42}(\vec{k},-\vec{k},\vec{k}_1,-\vec{k}_1) \tag{4.74}$$

For small k,

$$f^2(k) = 1 + 2\, f_2\, k^2 + 0(k^4) \tag{4.75}$$

where,

$$f_2 = \frac{1}{2}\frac{d^2}{dk^2}f(k)|_{k=0} = \frac{1}{2}[\beta_2 - 2a\gamma(0)] \tag{4.76}$$

We substitute for $u^*_{22}(0)$ from equation (4.68) in equation (4.69) and use equation (4.74), (4.75) and (4.76). This results in

$$- 2b_2a = 2\gamma(0)\ u^{*}_{21}{}^2(0)[\beta_2-2a\gamma(0)]$$

$$-2[2\gamma(0)\ u_{21}(0)\ C_4 + \{\beta_2\gamma(0) + \gamma_2\}\ u^{*}_{21}{}^2(0)]$$

$$+ \frac{1}{2}\int_{\vec{k}_1} \psi(k_k)\ \nabla_{\vec{k}}\cdot\nabla_{\vec{k}}\ \bar{u}^{*}_{42}(\vec{k},-\vec{k},\vec{k}_1,-\vec{k}_1)\Big|_{k = 0} \tag{4.77}$$

Also,

$$\frac{1}{2}\nabla_{\vec{k}}.\nabla_{\vec{k}}\ \bar{u}^{*}_{42}(\vec{k},-\vec{k},\vec{k}_1,-\vec{k}_1)\Big|_{\vec{k} = 0}$$

$$= 2A^2Bg_2g^2(k_1) - 4AC_1g(0) - \frac{1}{2}A^2\nabla_{\vec{k}}.\nabla_{\vec{k}}\ T(\vec{k}+\vec{k}_1) \tag{4.78}$$

where,

$$g_2 = \frac{1}{2}\frac{d^2}{dk^2}\ g(k)\Big|_{k = 0} = \frac{1}{4}\ [\gamma_2+2\{\beta_2-a\gamma(0)\}\gamma(0)] \tag{4.79}$$

Substituting from equations (4.78) and (4.79) in the equation (4.77),

$$- 2b_2a = - 2\ u^{*}_{21}{}^2(0)\ [\gamma_2 + 2a\gamma^2(0)] - 4\gamma(0)\ u^{*}_{21}(0)\ C_4$$

$$+ \frac{1}{2}A^2B^2[\gamma_2+2\{\beta_2-a\gamma(0)\}\gamma(0)] - 4ABC_1g(0)$$

$$- \frac{1}{6}A^2\int_{\vec{k}_1} \psi(k_1)\ \nabla_{\vec{k}}.\nabla_{\vec{k}}\ T(\vec{k}+\vec{k}_1) \tag{4.80}$$

With the help of equation (4.41), and

$$C_4 = \frac{1}{2}\frac{d^2}{dk^2}\ u^{*}_{21}(k)\Big|_{k = 0} = C_1 - \frac{1}{2}AB[\beta_2-2a\gamma(0)] \tag{4.81}$$

and

$$g(0) = \frac{1}{2}\gamma(0) \tag{4.82}$$

Equation (4.80) simplifies to

$$-2b_2a = - \frac{A^2}{2}\int_{\vec{k}_1} \psi(k_1)\nabla_{\vec{k}}.\nabla_{\vec{k}}\ T(\vec{k}+\vec{k}_1)\Big|_{\vec{k} = 0} \tag{4.83}$$

or,

$$-2b_2a = - \frac{A^2}{2}\int_{\vec{k}_1} \psi(k_1)\nabla_{\vec{k}}.\nabla_{\vec{k}_1}\ T(\vec{k}_1) \tag{4.84}$$

From the definitions of T(k), G(k) and g(k), it follows that

$$\nabla_{\vec{k}_2}\cdot\nabla_{\vec{k}_2} T(\vec{k}_2) = 2 \int_{\vec{k}_4} \psi(k_4)\, \nabla_{\vec{k}_2}\cdot\nabla_{\vec{k}_2} \int_o^1 \lambda^{-1} \frac{1}{F^2}$$

$$\times \int_o^F \lambda'\psi(\lambda')d\lambda'.d\lambda \qquad (4.85)$$

Let $I_1(\lambda)$ denote the quantity $\lambda\psi(\lambda)$. Then,

$$\nabla_{\vec{k}_2}\cdot\nabla_{\vec{k}_2} \int_o^1 \lambda^{-1} F^{-2} \int_o^F I_1(\lambda')d\lambda' d\lambda$$

$$= \int_o^1 \lambda^{-1}\{\int_o^F I_1(\lambda')d\lambda'(\frac{6\lambda^2}{F^4} - \frac{2\lambda^2}{F^4}(d-1)) - \frac{4\lambda^2}{F^3} I_1(F)$$

$$+ \frac{\lambda^2}{F^2} I_1'(F) + \frac{\lambda^2}{F^3} I_1(F)(d-1)\}d\lambda \qquad (4.86)$$

where the prime on $I_1'(F)$ denotes differentiation with respect to F. For d = 4, one obtains

$$\nabla_{\vec{k}_2}\cdot\nabla_{\vec{k}_2} \int_o^1 \lambda^{-1} F^{-2} \int_o^F I_1(\lambda')d\lambda'\, d\lambda$$

$$= \int_o^1 \lambda^{-1}\{\lambda^2 F^{-2} I_1'(F) - \lambda^2 F^{-3} I_1(F)\}d\lambda$$

$$= \int_o^1 \lambda F^{-1} \frac{d}{dF}\{\frac{I_1(F)}{F}\}d\lambda$$

$$= \int_o^1 \lambda \frac{1}{F}\frac{d}{dF}\psi(F)\, d\lambda \qquad (4.87)$$

From this, equation (4.72) follows immediately.

The solution for the fixed point function $u_{22}^*(k)$ may be obtained by the same procedure as we used to determine the other fixed point functions $u_{41}^*(\{\vec{k}_i\})$, $u_{21}^*(k)$, $u_{62}^*(\{\vec{k}_i\})$ and $u_{42}^*(\{\vec{k}_i\})$. We may choose $\xi = 0$ in equation (4.36) and put in the value of $u_{22}^*(0)$ given by the equation (4.68). The appropriate expressions for the coefficient functions $y(\lambda)$ and $z(\lambda)$ in equation (4.36) may be put in by comparing equation (4.35) with the equation (4.18). This procedure will determine $u_{22}^*(k)$. Since we do not use $u_{22}^*(k)$ to determine b to order ε^2, we do not carry out the explicit calculation.

PERTURBATIONS TO CRITICAL HAMILTONIANS NEAR FROM DIMENSION

5.1 Renormalization Group Equations for the n-Spin Perturbation Functions, $V_n(\vec{k}_1,\vec{k}_2\ldots,\vec{k}_n)$

The renormalization-trajectory of a slighly perturbed critical Hamiltonian does not, in general, terminate in a fixed point. However, we may assume that it passes by close to the fixed point associated with the original (unperturbed) critical Hamiltonian and in the neighborhood of the fixed point we may write,

$$H_t[S] = H^* + \sum_i O^i_m[S] \exp\{-d^i_m\, t\} \tag{5.1}$$

where $O^i_m[S]$ is an eigenfunction of the linear renormalization group operator and d^i_m the corresponding eigenvalue. The eigenvalue equation for d^i_m is obtained by substituting (5.1) in the equation (2.30) and keeping only terms to first order in $O^i_m[S]$:

$$\begin{aligned}-d^i_m\, O^i_m[S] &= \int_{\vec{k}} \{\frac{d}{2} S_{\vec{k}} + [b + \beta(k)]\, S_{\vec{k}} + \vec{k}.\nabla_{\vec{k}} S_{\vec{k}}\} \\ &\quad \times \frac{\delta O^i_m[S]}{\delta S_{\vec{k}}} + \int_{\vec{k}} [b + \beta(k)]\gamma(k)\ \{2\, \frac{\delta H^*}{\delta S_{-\vec{k}}} \\ &\quad \times \frac{\delta O^i_m}{\delta S_{\vec{k}}} + \frac{\delta^2 O^i_m}{\delta S_{\vec{k}} \delta S_{-\vec{k}}}\}\end{aligned} \tag{5.2}$$

The functional $O^i_m[S]$ is a localized functional of S. This means that if $O^i_m[S]$ is expanded in powers of S, then

$$\begin{aligned}O^i_m[S] &= \int_{\vec{k}_1} V^i_1(\vec{k}_1)\, S_{\vec{k}_i} + \frac{1}{2!} \int_{\vec{k}_1}\int_{\vec{k}_2} V^i_2(\vec{k}_1,\vec{k}_2)\, S_{\vec{k}_1} S_{\vec{k}_1} \\ &\quad + \frac{1}{3!} \int_{\vec{k}_1} \int_{\vec{k}_1} \int_{\vec{k}_3} V^i_3(\vec{k}_1,\vec{k}_2,\vec{k}_3) S_{\vec{k}_1} S_{\vec{k}_2} S_{\vec{k}_3} + \ldots\end{aligned} \tag{5.3}$$

where $V^i_1(\vec{k}_1)$, $V^i_2(\vec{k}_1, \vec{k}_2)$, $V^i_3(\vec{k}_1, \vec{k}_2, \vec{k}_3)$, etc., are analytic in $\vec{k}_1$, $\vec{k}_2$, and $\vec{k}_3$ for real values of the components of $\vec{k}_1$, $\vec{k}_2$, and $\vec{k}_3$. We substitute the expansion (5.3) in the equation (4.2) and obtain the following ordinary partial differential equations for the n-spin perturbation functions, $V_n^i(\vec{k}_1, \vec{k}_2,\ldots,\vec{k}_n)$:

$$-d_m^i \, V_1^i(\vec{k}_1) = A^*(\vec{k}_1)V_1^i(\vec{k}_1) + \int_{\vec{k}_2} [b + \beta(k_2)]\gamma(k_2) \times V_3^i(\vec{k}_1, \vec{k}_2, \vec{k}_3) \tag{5.4}$$

$$-d_m^i \, V_2^i(\vec{k}_1, \vec{k}_2) = [A^*(\vec{k}_1) + A^*(\vec{k}_2)] \, V_2^i(\vec{k}_1, \vec{k}_2) + \int_{\vec{k}_3} [b + \beta(k_3)]\gamma(k_3)V_4^i(\vec{k}_1, \vec{k}_2, \vec{k}_3, -\vec{k}_3) \tag{5.5}$$

$$-d_m^i \, V_3^i \, (\vec{k}_1, \vec{k}_2, \vec{k}_3) = [A^*(\vec{k}_1) + A^*(\vec{k}_2) + A^*(\vec{k}_3)] \times V_3^i(\vec{k}_1, \vec{k}_2, \vec{k}_3) - 2[b + \beta(\vec{k}_1+\vec{k}_2+\vec{k}_3)] \times \gamma(\vec{k}_1 + \vec{k}_2 + \vec{k}_3)V_1^i(\vec{k}_1 + \vec{k}_2 + \vec{k}_3) \times u_4^*(\vec{k}_1, \vec{k}_2, \vec{k}_3, -\vec{k}_1 -\vec{k}_2 -\vec{k}_3) \tag{5.6}$$

In general,

$$-d_m^i \, V_n^i(\vec{k}_1,\ldots,\vec{k}_n) = \sum_i A^*(\vec{k}_i) \, V_n^i(\vec{k}_1,\ldots,\vec{k}_n) - 2 \sum_P \sum_P [b + \beta(\vec{k}_1+\ldots + \vec{k}_{2p-1})]\gamma(\vec{k}_1+\ldots + k_{2p-1}) \times u_{2p}^*(\vec{k}_1,\ldots, \vec{k}_{2p-1}, \vec{k}) \, V_{n+2-2p} \, (\vec{k}_{2p},\ldots,\vec{k}_n, -\vec{k}) + \int_{\vec{k}_{n+1}} [b + \beta(\vec{k}_{n+1})]\gamma(\vec{k}_{n+1}) \, V_{n+2} \, (\vec{k}_1,\ldots, \vec{k}_n, \vec{k}_{n+1}, -\vec{k}_{n+1}) \tag{5.7}$$

where,

$$A^*(\vec{k}_1) = -\vec{k}_i \cdot \nabla_{\vec{k}_i} - \frac{1}{2} d + [b + \beta(k_i)][1-2\gamma(k_1)u_2^*(\vec{k}_i)] \qquad (5.8)$$

Σ_p denotes the sum over inequivalent permutations of $(\vec{k}_1,\ldots,\vec{k}_n)$ in the $u_{2p}^* \times V_{n+2-2p}$ term in equation (5.7) and $\vec{k} = -\vec{k}_1 \ldots \vec{k}_{2p-1}$.

We notice that the renormalization group equations for the function V^i fall into two disjoint classes, one containing odd V^i,s and one containing even V^i,s. Thus the eigenfunction $0_m^i[S]$, in equation (5.3) contains either odd powers of spin or even powers of spin but not both. We denote the two classes of the eigenfunctions by $0_m^1[S]$ and $0_m^2[S]$ respectively:

$$0_m^1[S] = \int_{\vec{k}} V_1(\vec{k}_1)\, S_{k_1} + \frac{1}{3!} \int_{\vec{k}_1} \int_{\vec{k}_2} \int_{\vec{k}_3} V_3(\vec{k}_1, \vec{k}_2, \vec{k}_3)$$

$$\times\, S_{\vec{k}_1}\, S_{\vec{k}_2}\, S_{\vec{k}_3} + \ldots \qquad (5.9)$$

$$0_m^2[S] = \frac{1}{2} \int_{\vec{k}_1} \int_{\vec{k}_2} V_2(\vec{k}_1, \vec{k}_2) S_{\vec{k}_1}\, S_{\vec{k}_2} + \frac{1}{4!} \int_{\vec{k}_1} \int_{\vec{k}_2} \int_{\vec{k}_3} \int_{\vec{k}_4}$$

$$V_4(\vec{k}_1, \vec{k}_2, \vec{k}_3, \vec{k}_4)\, S_{\vec{k}_1} S_{\vec{k}_2} S_{\vec{k}_3} S_{\vec{k}_4} + \ldots \qquad (5.10)$$

5.2 The Epsilon Expansion for the Relevant Eigenfunctions and Eigenvalues.

We have determined the fixed point interactions occurring in equation (5.2), (5.4)-(5.8), in $(4 - \varepsilon)$ dimensions, to order ε^2. It is therefore natural to seek eigenfunctions which are an analytic continuation of the eigenfunctions in four dimensions. This suggests the following epsilon expansions for the relevant odd and even eigenfunctions respectively:

$$d_m^1 = 1 + d_{m1}^1\, \varepsilon + d_{m2}^1\, \varepsilon^2 + \ldots \qquad (5.11)$$

$$d_m^2 = 2 + d_{m1}^2\, \varepsilon + d_{m2}^2\, \varepsilon^2 + \ldots \qquad (5.12)$$

$$V_1 = V_{10} + V_{11}\, \varepsilon + V_{12}\, \varepsilon^2 + \ldots \qquad (5.13)$$

$$V_3 = V_{31}\,\varepsilon + V_{32}\,\varepsilon^2 + \ldots \tag{5.14}$$

$$V_5 = V_{52}\,\varepsilon^2 + \ldots \tag{5.15}$$

$$V_2 = V_{20} + V_{21}\,\varepsilon + V_{22}\,\varepsilon^2 + \ldots \tag{5.16}$$

$$V_4 = V_{41}\,\varepsilon + V_{42}\,\varepsilon^2 + \ldots \tag{5.17}$$

$$V_6 = V_{62}\,\varepsilon^2 + \ldots \tag{5.18}$$

where $V_{10}(k_1) = f(k_1)$ and $V_{20}(k_1, k_2) = f(k_1)f(k_2)$ are the odd and even relevant Gaussian eigenfunctions respectively (Eq. 3.29 with $n = 0$ and Eq. 3.37). d^1_m and d^2_m are respectively the eigenvalues of the eigenfunctions $O^1_m[S]$ and $O^2_m[S]$ defined in equations (5.9) and (5.10). The epsilon expansions, equations (5.11)-(5.17), may be substituted in equations (5.4)-(5.7) to obtain a set of equations very similar to equations (4.17)-(4.21) which may be solved by the methods of section (4). In the case of the odd relevant eigenvector these methods may be circumvented by noting a general relation between the translationally invariant form of O^1_m and the fixed point Hamiltonian H^{*}[23]. It may be verified by direct substitution that

$$S_0 + \gamma(0)\,\frac{\delta H^*}{\delta S_0}$$

is an eigenfunction of the linearized renormalization group operator with the eigenvalue

$$d^1_m = \frac{d}{2} + b,$$

or, by equation (3.44)

$$\eta = 2(1-b) \tag{5.19}$$

Thus noting equation (4.7), (4.48), and (4.72)

$$\eta = -\frac{A^2\varepsilon^2}{a}\int_{\vec{k}_2}\psi(k_2)\int_{\vec{k}_4}\psi(k_4)\int_0^1 \lambda\, F^{-1}\,\frac{d}{dF}\,\psi(F)\,d\lambda + O(\varepsilon^3) \tag{5.20}$$

where $F = |\lambda\vec{k}_2 + \vec{k}_4|$

No such identity is evidently available for O^2_m. However, the methods of section (4) yield, to order ε

$$V_{41}(\{\vec{k}_i\}) = -\ 2A\prod_i f(k_i)\ \sum_P g(\vec{k}_1+\vec{k}_2+\vec{k}_3) \tag{5.21}$$

$$V_{21}(\{\vec{k}_i\}) = \pi_i f(k_i)\ [C_6 - AB\gamma(0) + C_1 \sum_i k_i^2 g(k_i) + AB \sum_i g(k_1) - 2AT(\vec{k}_1 + \vec{k}_2)] \tag{5.22}$$

where C_6 is an arbitrary constant.

and

$$d^2_{m1} = -1 + 4A \int_{\vec{k}_3} \psi(k_3)g(k_3) \tag{5.23}$$

Using equation (4.61), we have to order ε

$$d^1_m = 2 - \frac{2}{3}\varepsilon + O(\varepsilon^2) \tag{5.24}$$

Noting the relationship between the relevant eigenvalues of the linear renormalization group operator and the critical exponents, equations (3.39)-(3.45), we obtain

$$\alpha = \frac{1}{6}\varepsilon + O(\varepsilon^2)$$

$$\beta = \frac{1}{2} - \frac{1}{6}\varepsilon + O(\varepsilon^2)$$

$$\gamma = 1 + \frac{1}{6}\varepsilon + O(\varepsilon^2)$$

$$\delta = 3 + \varepsilon + O(\varepsilon^2)$$

$$\Delta = \frac{3}{2} + O(\varepsilon^2)$$

$$\nu = \frac{1}{2} + \frac{1}{12}\varepsilon + O(\varepsilon^2)$$

UNIVERSALITY OF THE EXPONENT η TO ORDER ε^2

The expression for the η to order ε^2 obtained in the preceding section apparently depends on a, $\beta(k)$ and $\gamma(k)$. If this were true, this would bring into serious question the concept of the universality of critical exponents. We show in this section that this is not the case. The coefficient of ε^2 in the equation (5.71) is universal independent of the parameter a and the cutoff functions $\beta(k)$ and $\gamma(k)$.[24] Our analysis confirms a recent numerical computation by Golner and Riedel[25] who evaluated our expression for η to order ε^2 by computer for the choices

$$\gamma(k) = 1, \ \beta(k) = 2k^2; \text{ and,}$$

$$\gamma(k) = 1, \ \beta(k) = k^4;$$

and obtain $\frac{1}{54}$ in both cases.

6.1 An Integral Identity for the Function $\psi(k)$

We start with the observation that the function $\psi(k)$ can be written as

$$\psi(k) = -\frac{1}{2}\frac{1}{k}\frac{d}{dk}\left[\frac{1}{\phi(k)}\right] \tag{6.1}$$

where,

$$\phi(k) = [1 + 2a \int_o^k [1 + \beta(k')]\gamma(k')\, k' \times \exp\{2 \int_o^{k'} k''^{-1}\, \beta(k'')\, dk''\} \tag{6.2}$$

The function $\phi(k)$ is a monotonically increasing function of k and consequently equation (6.1) yields the identity

$$\int_o^\infty k\psi(k)dk = \frac{1}{2a} \tag{6.3}$$

6.2 Reduction of the Expression for η to a Threefold Integral Over k_2, k_4 and F

With the help of the equation (6.3) the expression for A in equation (4.61) can be evaluated easily. We obtain

$$A = \frac{a^2}{3\pi} \tag{6.4}$$

Substituting the above value for A in the equation (5.20) and writing the four dimensional integrals explicitly, we obtain up to order ε^3

$$\eta = -\frac{8a^3\varepsilon^2}{9\pi} \int_o^\infty k_2\psi(k_2)dk_2 \int_o^\infty k_4\psi(k_4)dk_4 I(k_2,k_2,F) \tag{6.5}$$

where,

$$I(k_2,k_4,F) = k_2^2\, k_4^2 \int_o^\pi \sin^2\Theta d\Theta \int_o^1 \lambda d\lambda\ F^{-1} \frac{d}{dF}\psi(F) \tag{6.6}$$

and θ is the angle between vectors $\vec{k}_2$ and $\vec{k}_4$. Making use of the relation

$$\frac{d\lambda}{F} = \frac{dF}{(\lambda k_2^2 + k_2 k_4 \cos\,\theta)} \tag{6.7}$$

we can write (6.6) as a Stieltjes integral

$$I = k_2\, k_4^2 \int_o^{\pi} \sin^2\theta d\theta \int_{\lambda=0}^{1} \lambda(\lambda k_2 + k_4 \cos\theta)^{-1}$$

$$x\ d[\psi(F(\lambda)) - \psi(k_4 \sin\theta)] \tag{6.8}$$

Here the term $\psi(k_4\ \sin\theta)$ has been subtracted to make the integrand finite when $\lambda k_2 + k_4\cos\theta$ vanishes. Integration by parts yields

$$I = \int_{|k_2-k_4|}^{(k_2+k_4)} F\psi(F)\tan\,\alpha\ dF - k_2 \int_o^{k_4} F\psi(F)(k_4^2-F^2)^{-\frac{1}{2}}$$

$$x\ (\cos\,\alpha)^{-1}\ dF + J \tag{6.9}$$

where α is the angle between k_2 and F in a triangle whose sides are k_2, F and k_4, and

$$J = -\ k_2\, k_4^3 \int_o^{\pi} \sin^2\theta\ \cos\theta\ d\theta \int_o^1 \frac{[\psi(F) - \psi(k_4\ \sin\theta)]}{(\lambda k_2 + k_4\cos\theta)^2} d\lambda \tag{6.10}$$

J can be evaluated easily by changing the variables of integration in equation (6.10) to F and F', where

$$F' = k_4\ \sin\,\theta \tag{6.11}$$

The change from λ to F is accomplished with the help of equation (6.7). The domain of integration depends on whether $k_2 > k_4$ or $k_2 < k_4$. If $k_2 > k_4$, we obtain

$$J = -\ k_2\ k_4 \int_o^{\pi/2} F'^2\ \cos\theta d\theta \int_{k_4}^{F^+} \frac{[\psi(F)-\psi(F')]}{(F^2-F'^2)^{3/2}}\ FdF$$

$$-k_2\ k_4 \int_{\pi/2}^{\pi} F'^2\ \cos\theta d\theta\ \{- \int_{k_4}^{F'} \frac{[\psi(F)-\psi(F')]}{(F^2-F'^2)^{3/2}}\ FdF$$

$$+ \int_{F'}^{F^-} \frac{[\psi(F)-\psi(F')]}{(F^2-F'^2)^{3/2}} FdF\} \tag{6.12}$$

where

$$F^+ = [(k_2 + \sqrt{k_4^2-F'^2})^2 + F'^2]^{\frac{1}{2}} \tag{6.13}$$

and

$$F^- = [(k_2 - \sqrt{k_4^2-F'^2})^2 + F'^2]^{\frac{1}{2}} \tag{6.14}$$

If $k_2 < k_4$,

$$J = -k_2 k_4 \int_0^{\pi/2} F'^2 \cos\theta d\theta \int_{k_4}^{F^+} \frac{[\psi(F)-\psi(F')]}{(F^2-F'^2)^{3/2}} FdF$$

$$- k_2 \; k_4 \int_{\pi/2}^{\theta_o} F'^2 \cos\theta d\theta \{-\int_{k_4}^{F'} \frac{[\psi(F)-\psi(F')]}{(F^2-F'^2)^{3/2}} FdF$$

$$+ \int_{F'}^{F^-} \frac{[\psi(F)-\psi(F')]}{(F^2-F'^2)^{3/2}} FdF\}$$

$$+ k_2 k_4 \int_{\theta_o}^{\pi} F'^2 \cos\theta d\theta \int_{k_4}^{F^-} \frac{[\psi(F)-\psi(F')]}{(F^2-F'^2)^{3/2}} FdF \tag{6.15}$$

where

$$\theta_o = \cos^{-1}(-\frac{k_2}{k_4}) \tag{6.16}$$

The radical $(F^2 - F'^2)^{\frac{1}{2}}$ in equations (6.12) and (6.15) comes from the factor $\lambda k_2 + k_4 \cos\theta$ in the equation (6.10). When $\cos\theta$ is negative, the expression $\lambda k_2 + k_4 \cos\theta$ is also negative along the path from $F = k_4$ to $F = F'$. We have, however, arranged the signs in equations (6.12) and (6.15) such that radical $(F^2 - F'^2)^{\frac{1}{2}}$ is always to be taken positive. The domain of integration of the $F - \theta$ plane can be understood with the help of figures (1a), (1b) and (1c). When $\cos\theta$ is positive. F monotonically increases from its value k_4 corresponding to $\lambda = 0$ to its value F^+ corresponding to $\lambda = 1$. When $\cos\theta$ is negative, F similarly increases from k_4 to F^-. However, in this case, the increase is not monotonic. F first reaches its minimum value F' corresponding to the case when $\lambda k_2 + k_4 \cos\theta$ is zero and

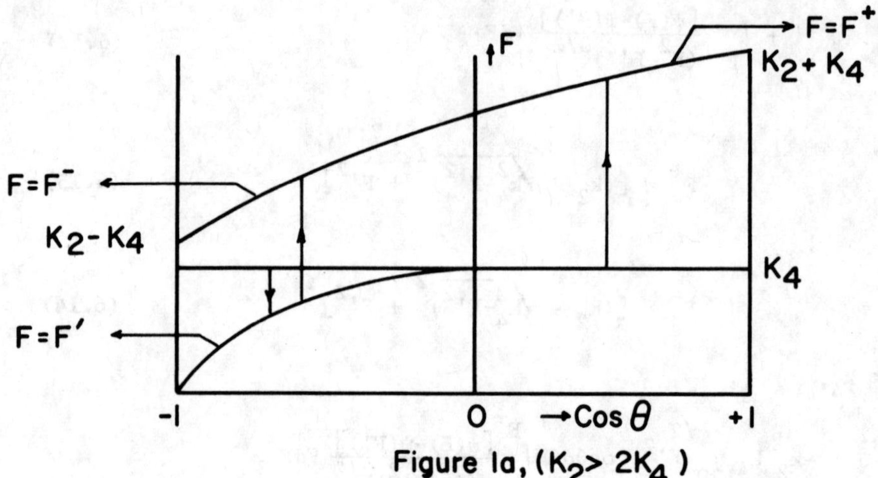

Figure Ia, ($K_2 > 2K_4$)

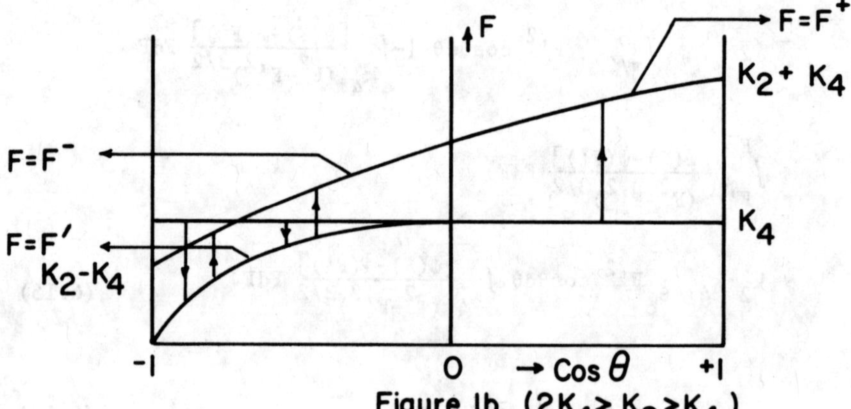

Figure Ib, ($2K_4 > K_2 > K_4$)

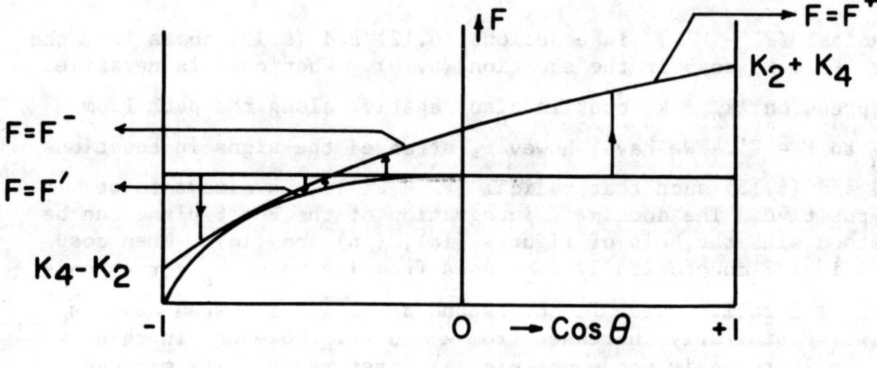

Figure Ic, ($K_2 < K_4$)

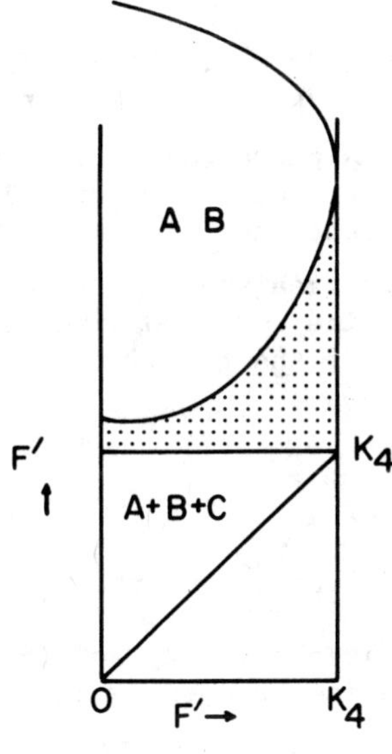

Figure 2a, ($K_2 > 2K_4$)

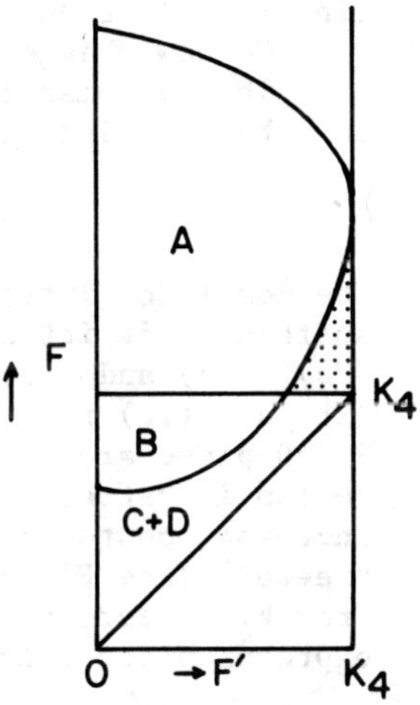

Figure 2b, ($2K_4 > K_2 > K_4$)

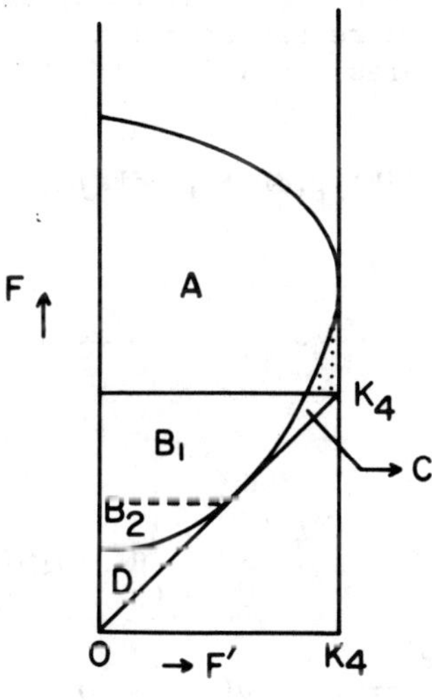

Figure 2c, ($K_2 < K_4$)

then rises to the value F^- corresponding to $\lambda = 1$. Figure (1a) depicts the situation when $k_2 > 2k_4$. When $2k_4 > k_2 > k_4$, the domain of integration is shown in figure (1b). Algebraically, the two cases depicted in figures (1a) and (1b) are not different. The domain of integration is, however, different when $k_2 < k_4$. The source of this difference is that in this case the line F^- corresponding to $\lambda = 1$ touches the parabolic line corresponding to $\lambda k_2 + k_4 \cos\theta = 0$. The point of contact of the two lines is at $\cos\theta = \cos\theta_o$ and

$$F = F_o = (k_4^2 - k_2^2)^{\frac{1}{2}} \tag{6.17}$$

When $\cos\theta < \cos\theta_o$ the region between the line F^- and the parabolic line $F = F'$ is not traversed.

We now change the variable θ in equations (6.12) and (6.15) to F'. The Jacobian of this transformation is obtained by differentiating the equation (6.11);

$$dF' = k_4 \cos\theta d\theta \tag{6.18}$$

The domain of integration over $F - F'$ plane becomes a three-sheeted surface as depicted in figures (2a), (2b) and (2c). The figures (2a), (2b) and (2c) on $F - F'$ plane correspond to the figures (1a), (1b) and (1c) respectively on the $F - \theta$ plane. The surfaces on the $F - \theta$ plane are two-sheeted since, as explained earlier, part of the region is retraversed back as λ in equation (6.10) continues to increase monotonically. The surfaces of $F - F'$ plane are three sheeted since F' in equation (6.11) varies from zero to k_4 and back from k_4 to zero as $\cos\theta$ monotonically increases from 1 to -1. The expression for J in terms of the variables F and F' is: If $k_2 > k_4$,

$$J = \left[-\int_o^{k_4} dF' \int_{k_4}^{F^+} dF + \int_o^{k_4} dF' \int_{F'}^{k_4} dF + \int_o^{k_4} dF' \int_{F'}^{F^-} dF\right]$$

$$\times \frac{F'^2 F}{(F^2 - F'^2)^{3/2}} \left[\psi(F) - \psi(F')\right] \tag{6.19}$$

and, if $k_2 < k_4$,

$$J = \left[-\int_o^{k_4} dF' \int_{k_4}^{F^+} dF + \int_{F_o}^{k_4} dF' \int_{F'}^{k_4} dF + \int_{F_o}^{k_4} dF' \int_{F'}^{F^-} dF\right.$$

$$\left. + \int_o^{F_o} dF' \int_{F^-}^{k_4} dF\right] \frac{F'^2 F}{(F^2 - F'^2)^{3/2}} \left[\psi(F) - \psi(F')\right] \tag{6.20}$$

The radical $(F^2 - F'^2)^{\frac{1}{2}}$ in equations (6.19) and (6.20) is again to be taken with positive sign. We note that the shaded regions in figures (2a), (2b) and (2c) contribute zero to J since these regions are traversed twice and each time the contribution from each to J is the same in magnitude but of opposite sign. The region D in figure (2c) is not traversed. The contribution to J can be simply written as

$$J = - J_{A+B} + 2\ J_{B+C+D}$$

$$\text{for } k_2 > k_4 \qquad \text{(figures 2a and 2b)} \tag{6.21}$$

and

$$J = - J_{A+B_1+B_2} + 2\ J_{B_1+B_2+C}$$

$$\text{for } k_2 < k_4 \qquad \text{(figure 2c)} \tag{6.22}$$

where J_{A+B} denotes the integral of the function

$$\frac{F'^2\ F}{(F^2-F'^2)^{3/2}} [\psi(F)-\psi(F')]$$

over the area A + B, etc. In equations (6.19) and (6.20), the integrals involving $\psi(F')$ can be reduced to one-fold integrals over F' by integrating over F. The integrals involving $\psi(F)$ can also be reduced to one-fold integral over F by changing the order of integration and integrating F' first. Carrying out this integral involves the evaluation of the principal value of the function arc sin α. When $k_2 > k_4$, the angle α is obtuse in the region B_2 in figure (2c) and there the principal value of arc sin α is $(\pi - \alpha)$. We obtain if $k_2 > k_4$

$$J = \int_{k_2-k_4}^{k_2+k_4} F\psi(F)\ \alpha\ dF - \pi \int_{o}^{k_4} F\ (F)dF - K \tag{6.23}$$

and, if $k_2 < k_4$

$$J = \int_{k_4-k_2}^{k_2+k_4} F\psi(F)\ \alpha\ dF - k_2 \int_{k_4-k_2}^{k_4} F\psi(F)dF - K \tag{6.24}$$

where

$$K = \int_{|k_2-k_4|}^{k_2+k_4} F\psi(F)\ \tan\alpha\ dF - k_2 \int_o^{k_4} \frac{F\psi(F)dF}{(k_4^2-F^2)^{\frac{1}{2}}\cos\alpha} \tag{6.25}$$

Substituting equations (6.23), (6.24) and (6.25) in the equation (6.9), we get:

$$I = \int_{k_2-k_4}^{k_2+k_4} F\psi(F)\alpha dF - \pi \int_o^{k_4} F\psi(F)dF \quad \text{for } k_2 > k_4 \tag{6.26}$$

and

$$I = \int_{k_4-k_2}^{k_2+k_4} F\psi(F)\ \alpha\ dF - \pi \int_{k_4-k_2}^{k_4} F\psi(F)dF \quad \text{for } k_2 < k_4 \tag{6.27}$$

6.3 A Geometric Evaluation of the Threefold Integral for η.

The significance of the result arrived at in equations (6.26) and (6.27) can be understood easily with the help of figure (3a). In this figure, the three orthogonal axes denote the lengths of the vectors $\vec{k}_2$, $\vec{k}_4$ and $\vec{F}$ respectively where

$$\vec{F} = \lambda\vec{k}_2 + \vec{k}_4 \tag{6.28}$$

We consider a triangle in the k_2, k_4, F space which passes through the points (C,0,0), (0,C,0), and (0,0,C) where C is an arbitrary constant. In figure (3b), this triangle has been drawn on a two dimensional diagram. The interior equilateral triangle in figure (3b) is obtained by joining the mid-points of the sides of the larger triangle. The equations of the lines joining the mid-points are indicated on the figure. The shaded portions in figure (3b) show the domaining of the integrals in equations (6.26) and (6.27). The integrand for the region shaded with parallel lines is $\alpha F\psi(F)$ and for the region shaded with dots, it is $-\pi F\psi(F)$. The integrand for the region shaded with both parallel lines and dots is $\alpha F\psi(F) - \pi F\psi(F)$. The large equilateral triangle in figure (3b), in a sense, denote the whole k_2, k_4, F space since all possible values of these variables are incorporated by contracting or dilating the triangle by a scale factor. The reason that the domain of integration on this triangle is asymmetrical is that asymmetrical roles have been assigned to the variables k_2, k_4, and F in our analysis. A little reflection shows that we can put k_2, k_4, and F on equal footing by taking the sum of the six permutations of k_2, k_4 and F in equations (6.5), (6.26) and (6.27) and dividing by six. We can write the equation (6.5) as

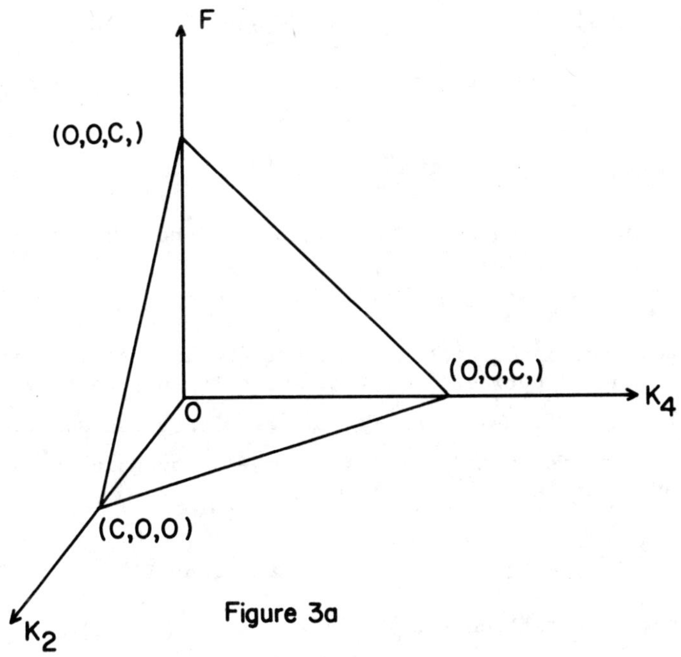

Figure 3a

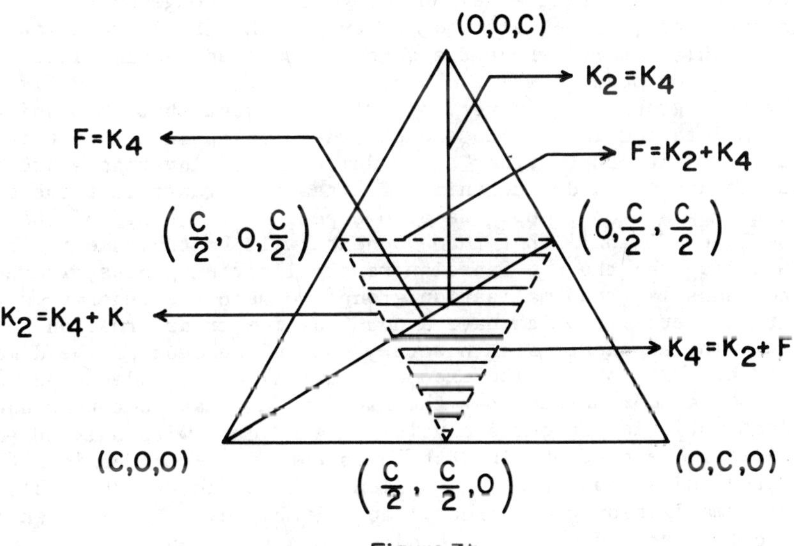

Figure 3b

$$\eta = -\frac{8a^3\varepsilon^2}{9\pi} \int k_2\psi(k_2)dk_2 \int k_4\psi(k_4)dk_4 \int F\psi(F)dF\ L$$

where integrals are over the interior of the larger triangle in figure (3b);

$$L = \frac{1}{6}\left[2(\alpha+\beta+\gamma) - 3\pi\right] \tag{6.30}$$

if the point of integration is inside the inner triangle;

$$L = \frac{1}{6}(-\pi) \tag{6.31}$$

if the point of integration is outside the inner triangle; and β and γ are the angles between k_2 and k_4 and between k_4 and F, respectively, in the triangle formed by k_2, k_4 and F. Since the sum of three angles of a triangle is π, in fact $L = -\frac{\pi}{6}$ both inside and outside the inner triangle. Thus we obtain

$$\eta = \frac{8a^3\varepsilon^2}{54} \int_o^\infty k_2\psi(k_2)dk_2 \int_o^\infty k_4\psi(k_4)dk_4 \int_o^\infty F\psi(F)dF \tag{6.32}$$

Taking note of equation (6.3),

$$\eta = \frac{\varepsilon^2}{54} \tag{6.33}$$

CONCLUDING REMARKS

It is seen from our analysis that, to the order in epsilon examined, the critical exponents given by the general Gaussian renormalization group are independent of the cutoff functions $\beta(k)$ and $\gamma(k)$ which characterize the general Gaussian renormalization group and the constant a which characterizes the line of the fixed points of this group. It is very natural to expect that this universality is maintained in all subsequent orders in epsilon. In this connection we note that Wegner[26] has demonstrated invariance properties of the exponents under certain infinitesimal transformations of the renormalization group. We also expect on the basis of the paper of Bell and Wilson[11] that the exponents will be the same for both the Gaussian and the non-Gaussian renormalization groups, where by the non-Gaussian renormalization groups we mean the renormalization group operators which have a non-Gaussian transformation kernel.

The exponents are, however, not independent of the dimensionality of the system. The exponents are also not independent of the number of components that the field $S_{\vec{k}}$ may have but this question does not arise in our analysis since we deal with only one-component fields. We have obtained the exponents in $4-\varepsilon$ dimensions to the first and second order in epsilon, but, of course, the value of the renormalization group theory can only be fully judged when its predictions are known in three dimensions and compared with experiment.

The epsilon expansion has been carried out to the fourth order in epsilon but it appears to be an asymptotic expansion. It is seen that if we set $\varepsilon = 1$ in the epsilon expansion expressions for the critical exponents, the first two or three terms yield results which compare fairly well with the high-temperature expansion results for the three dimensional Ising model as well as experiment.[9,27] There is, of course, no strong justification why the first few terms in the epsilon expansion should give "good" results for three dimensions when epsilon is set equal to unity but this practice is currently very popular. We may expect that in future there will be increased emphasis on finding alternative methods of solving the renormalization group equations in three dimension directly.

ACKNOWLEDGEMENTS

M. S. Green gratefully acknowledges the support of the J. S. Guggenheim Memorial Foundation and P. Shukla, of a Temple University Fellowship. The authors wish to acknowledge the generous hospitality of the Racah Institute of Physics of the Hebrew University extended under difficult circumstances. This work was supported by the National Science Foundation through Grant # GP41251X. It was submitted as a thesis to Temple University by P. Shukla in partial fulfillment of the requirements for the degree of philosophy.

REFERENCES

1. For a brief and very readable exposition to the problem of critical phenomena, see B. J. Josephson in Methods and Problems of Theoretical Physics, edited by J. E. Bowcock (North-Holland Publishing Company, 1970). For a more detailed account, see References 2-5.
2. M. E. Fisher, Repts. Prog. Phys. 30, 615 (1967).
3. L. P. Kadanoff et al., Rev. Mod. Phys. 39, 395 (1967).
4. Critical Phenomena: Proceedings of the International School of Physics, "Enrico Fermi", Varenna, 1970, Number 51, edited by M. S. Green (Academic Press, New York, 1971).
5. H. E. Stanley, Introduction to Phase Transitions and Critical Phenomena (Oxford University Press, New York, 1971).
6. See J.M.H. Levelt-Sengers, in the Proceedings of the I.U.P.A.P. Van der Waals Centennial Conference on Statistical Mechanics, Amsterdam, 1973, and, in Reference 7.
7. Renormalization Group in Critical Phenomena and Quantum Field Theory: Proceedings of a Conference, Philadelphia, 1973, edited by J. D. Gunton and M. S. Green (Department of Physics, Temple University.)
8. K. G. Wilson, Phys. Rev. B 4, 3174, 3184 (1971).
9. K. G. Wilson, as reported in K. G. Wilson and J. Kogut, Phys. Rep. 12 C, 76 (1974).
10. (a) F. J. Wegner, Phys. Rev. B 5, 4528 (1972).
    (b) F. J. Wegner, Phys. Rev. B 6, 1891 (1972).
11. T. L. Bell and K. G. Wilson, Phys. Rev. B 10, 3935 (1974).

12. K. G. Wilson and M. E. Fisher, Phys. Rev. Lett. 28, 248 (1972).
13. (a) P. Shukla and M. S. Green, Phys. Rev. Lett. 33, 1263 (1974).
(b) P. Shukla and M. S. Green, Phys. Rev. Lett. 34, 436 (1975).
14. I. M. Gel'fand and G. E. Shivol, Generalized Functions (Academic Press, New York, 1964), Vol. I.
15. See S. Chandrasekhar, Rev. Mod. Phys. 15, 1 (1943). The original papers are: A. D. Fokker, Ann. Physik 43, 812 (1914); M. Planck, Sitzungsber. Preuss. Akad, Wiss., (1917).
16. F. Wegner and A. Houghton, Phys. Rev. A 8, 401 (1973).
17. H. Cramer, Random Variables and Probability Distributions (Cambridge University Press, London, 1937).
18. M. E. Fisher, Rev. Mod. Phys. 46, 597 (1974).
19. L. D. Landau, Phys. Zurn. Sowjet-union 11, 26, 545 (1937). English translation: D. ter Haar, Men of Physics: L. D. Landau, Vol. II (Pergamon Press, Oxford, 1969).
20. L. S. Ornstein and F. Zernike, Proc. Sect. Sci, K. Med. Akad. Wet. 17, 793 (1914).
21. (a) M. E. Fisher and D. S. Gaunt, Phys. Rev. 133, A 224 (1964).
(b) B. Widom, Mol. Phys. 25, 657 (1973).
(c) G.'t Hooft and M. Veltman, Nucl. Phys. B 44, 189 (1972).
(d) K. G. Wilson, Phys. Rev. D 7, 2911 (1973).
22. R. Courant and D. Hilbert, Methods of Mathematical Physics (Interscience, New York, 1953), Vol. II.
23. F. J. Wegner in Phase Transitions and Critical Phenomena (Academic Press, New York, 1975), Vol. 6.
24. For an alternate analysis see J. Rudnick, Phys. Rev. Lett. 34, 438 (1975).
25. G. R. Golner and E. K. Riedel, Phys. Rev. Lett. 34, 171 (1975).
26. F. J. Wegner, J. Phys. C: Solid St. Phys. 7, 2098 (1974).
27. See M. Wortis in Reference 7.

Dr. Green is Professor of Physics at Temple University and is a fellow of the American Physical Society. He was Chief of the Statistical Physics Section at the National Bureau of Standards and has worked on fundamental problems in the statistical mechanics of time dependent phenomena. His recent work concerns the theory of critical phenomena and the renormalization group. He first met Julius when they were graduate students at Princeton. They were contemporaries in the Statistical Physics Section at the National Bureau of Standards.

Dr. Shukla is a research associate in the Department of Physics at Oxford University and this article describes part of his thesis work at Temple University. Currently he is working on magnetism in amorphous systems.

ON A NONLINEAR PERTURBATION THEORY WITHOUT SECULAR TERMS

II. CARLEMAN EMBEDDING OF NONLINEAR EQUATIONS IN AN INFINITE SET OF LINEAR ONES

Elliott W. Montroll*
Physical Dynamics, Inc., La Jolla, California 92037

Robert H. G. Helleman
Institute for Fundamental Studies
Department of Physics and Astronomy
University of Rochester, Rochester, New York 14627

ABSTRACT

C. Eminhizer and the authors[1] have recently developed iterative methods for solving the dynamics of extremely nonlinear anharmonic oscillators. These methods are re-examined here in a formalism proposed by Carleman[2] for the embedding of nonlinear dynamical equations in an infinite set of linear equations.

The methods are based on recursion formulae for the Fourier coefficients of Fourier series solutions of the dynamical equations and for coefficients of expansions of Fourier frequencies.

The series solutions avoid "Secular Terms" and "Small Denominators" and converge rapidly in large regimes of highly nonlinear behaviour.

I. INTRODUCTION

With few exceptions the complex mathematical problems encountered in statistical mechanics, fluid dynamics, and many body processes, can be investigated only through approximation procedures. The basic dynamics which generate these processes are frequently nonlinear. Two of the most commonly used methods in the theories of turbulence and many body physics are: (a) construction of a hierarchy of differential equations which relate correlation or Greens functions of successively higher order (the order being determined by the number of variables, say velocities, positions, spins, etc. which appear in the functions. (b) development of a perturbation expansion, often through the aid of diagrams, in powers of some "small" parameter or coupling constant (which in the actual range of interest might not be so small).

A serious difficulty with hierarchy formalisms is that uncontrolled closure hypotheses, which relate some higher order function to lower order ones, are made to terminate the hierarchy and replace it by one or two equations. The regime of validity of a closure hypothesis is generally very small and no estimation of errors can

*Permanent Address, Institute for Fundamental Studies, University of Rochester, Rochester, New York 14627

be made. The perturbation formalisms referred to in (b) are frought with divergence problems. The coefficients of various powers of the "small" parameters may be divergent integrals, or the series itself may have a small (or even zero) radius of convergence. While patching a theory by the summing of certain subclasses of diagrams before performing integrations, or by using the method of Padé approximants, has had some success, it is desirable to develop controlled perturbation expansions with larger radii of convergence which put more physical insight into the calculations.

The mathematicians and celestial mechanicians of the 19th century encountered similar perturbation theory difficulties in classical mechanics, especially as applied to three and many body orbit calculations. Considerable insight in perturbation theory can be obtained from the works of Poincaré and other late 19th and early 20th century mathematicians. Among the generical annoyances of divergences and slow convergence of series, two specific technical matters, the so called "Secular Terms" and of "Small Denominators" were nuisances which had to be coped with.

Since the detailed analysis presented in this paper will be made on the dynamics of (i) an anharmonic oscillator whose displacement $\phi(t)$ is governed by the equation

$$\ddot{\phi} + \omega_o^2\phi = R\phi^3 \tag{1}$$

and (ii) a pair of coupled anharmonic oscillators with displacements $\phi(\mu)$ governed by

$$\ddot{\phi}(\mu)+\omega_o^2(\mu)\phi(\mu) = R \sum_{k=0}^{3} a_{\mu k}[\phi(1)]^{3-k}[\phi(2)]^k \text{ , } \mu=1,2, \tag{2}$$

we base our summary of the classical difficulties on these two equations.

A popular style in perturbation theory is to convert a differential equation into an integral equation and iterate. Eq. (1) with initial conditions $\phi(0)=\phi_o$ and $\dot{\phi}(0)=z$ is equivalent to

$$\phi(t)=\phi_o(t) + (R/\omega_o) \int_0^t \phi^3(\tau) \sin[\omega_o(t-\tau)]d\tau , \tag{3}$$

where

$$\phi_o(t) = \phi_o \cos(\omega_o t) + (z/\omega_o)\sin(\omega_o t). \tag{4}$$

I.1 The Problem of the Secular Terms

Upon iteration one obtains a series expansion for $\phi(t)$ in powers of R. Under the simple initial condition z=0

$$\phi(t)/\phi_o = \cos\omega_o t + (R\phi_o^2/\omega_o) \int_0^t d\tau \sin\omega_o(t-\tau)[\cos^3\omega_o\tau + R(\ldots)+\ldots] \tag{5}$$
$$= \cos\omega_o t + (R\phi_o^2/\omega_o)[\tfrac{3}{4}\, t \sin\omega_o t + \text{periodic terms}]+R^2\{\ \}+\ldots$$

This first order *Secular Term*, proportional to t $\sin(\omega_o t)$, does, with increasing t, become arbitrarily large. Thus this approximation violates the conservation of energy condition which exists for an oscillator characterized by (1). The concept of *"Renormalization"* was first introduced into physics by Stokes[3] to respond to such situations. Let the harmonic frequency ω_o be renormalized, i.e. replaced by the "true" frequency of oscillation

$$\nu = \omega_o + R\omega_1 + R^2\omega_2 + \ldots, \tag{6}$$

where the ω-coefficients can be chosen so that no secular terms appear in (5). Then

$$\cos\omega_o t = \cos(\nu - R\omega_1 - R^2\omega_2 \ldots)t$$
$$= \cos\nu t + R\omega_1 t \sin\nu t + R^2\{\ \}+ \ldots$$

Eq. (5) becomes, after choosing

$$\omega_1 = -(3/4)\phi_o^2/\omega_o, \tag{7}$$

$$\phi(t)/\phi_o = \cos\nu t + (R\phi_o^2/32\omega_o^2)(\cos\nu t - \cos 3\nu t) + R^2\{\ \}+\ldots, \tag{8}$$

so that $\phi(t)$ no longer has secular terms to first order. A proper choice of coefficients can, in principle, be made so that *no secular terms appear in any order*.

Stokes[3] introduced the idea of renormalization but it remained for Lindstedt[4] and Poincaré[5] to develop a more systematic perturbation procedure to eliminate secular terms to *all* orders. The LP-method involves the construction of one of oldest hierarchies of differential equations used in dynamics. In the LP scheme one sets

$$\phi(t) = \phi_o(t) + R\phi_1(t) + R^2\phi_2(t)+\ldots \tag{9}$$

Then, for $\phi_j(t)$, a *linear* differential equation is derived with an inhomogeneous right hand side which is a nonlinear function of the lower order ϕ_k's, with k<j, which are presumed to be known. The hierarchy is arranged so that at each order a new ω_j and $\phi_j(t)$ is obtained. Unfortunately, their recursion procedure for the determination of new functions ϕ_j becomes so cumbersome that one seldom has the courage to proceed beyond second or third order.

In reference 6 we developed a new recursion scheme, solely in terms of the, discrete, Fourier coefficients of the Fourier series expansion of $\phi(t)$ and the new frequency coefficients of eqs. (6-8). Having numbers rather than functions in our recursion relations also allowed us to take advantage of a technology not available to Poincaré and Lindstedt, the high speed electronic computer and, in less than a minute of computer time, one can go as high as 20-th order. Unfortunately, the particular expansion of ν about the *harmonic* frequency ω_o in (6) produces a radius of convergence for the perturbation series which is only

$$|\theta|<1 \quad , \quad \text{with } \theta \equiv R\phi_o^2/\omega_o^2 \ . \tag{10}$$

However, in reference 1, C. Eminhizer and the authors have produced formalisms which do converge for all periodic solutions, i.e. $-\infty < \theta < +1$. Their rate of convergence is extremely rapid[1], as we will demonstrate in section I.3, using eq. (27).

I.2 The Problem of the Small Denominators

In addition to these practical advantages, the formalism of reference 1 avoids another classical difficulty, the problem of the *"Small Denominators (or Divisors)"*, known to prevent convergence in series solutions of systems of coupled equations. In eq. (2) we have a system of two coupled anharmonic oscillators. When we compare the *system* (2) with the *single* anharmonic oscillator (1) we see that the equation with $\ddot{\phi}(1)$ is the same as for the single oscillator (1), but "driven" by terms containing $\phi(2)$. One effect this has is to introduce periodic "driving" terms with frequencies different from the one appearing in the solution of the single anharmonic oscillator (1) [i.e. the ν of (6)]. Let us consider the simplest possible example of this effect

$$\ddot{\phi} + \omega_o^2\phi = R\phi^3 + B\cos\gamma t, \qquad \text{with } \dot{\phi}(0)=0, \tag{11}$$

a *driven* anharmonic oscillator. If we construct the equivalent integral equation, cf. (3), we will now, upon iteration [cf. (4) and (5)], obtain periodic terms with frequencies of the form $(n\nu + p\gamma)$, where ν is again chosen so as to avoid secular terms, cf. (6), and γ is the frequency of the driving term in (11) [The ω_o in the kernel $\sin \omega_o(t-\tau)$ of the integral equations (5) and (3) is also renormalized to ν]. Thus, after several iterations, one has among the many terms to be integrated, terms of the form

$$\int_0^t d\tau \ \sin\nu(t-\tau)\{[\cos((n-1)\nu-p\gamma)\tau]+..= \{ \frac{-1+\cos(n\nu-p\gamma)t}{(n\nu-p\gamma)} + \ ... \ +.., \tag{12}$$

where n and p are integers. The denominator $(n\nu-p\gamma)$ can become arbitrary small for suitable values of n and p. Hence the series diverges in general and is useless even as an asymptotic approximation.

We have overcome this problem for the case where ν *and* γ *are rationally dependent*, or "commensurate", i.e. when there exist integers m_1 and m_2 such that

$$m_2\nu - m_1\gamma = 0 \qquad (m_2 \text{ and } m_1 \text{ relative primes}); \text{ hence} \tag{13}$$

$$\sigma \equiv \gamma/\nu = m_2/m_1 \tag{14}$$

is a rational number. While this entails a theoretical restriction, rational values of σ do suffice for most of our practical calculations[1], much in the same way as the rational numbers suffice for most practical computations; witness the success of the digital computer. Note that in the usual series, and in the one represented by (12), the case with σ rational yields the worst possible divergencies. For this reason most of the theoretical efforts in the literature were concentrated on the case with σ *ir*rational, cf. the references in ref. 1.

When (13) is satisfied the $[n\nu + p\gamma]$ are no longer independent but are in fact all multiples of the "greatest common frequency" ν_r, with

$$\nu_r \equiv \gamma/m_2 = \nu/m_1, \qquad (m_2 \neq m_1) \tag{15}$$

which we call the *"Recurrence Frequency"*. The integral equation version of (11) becomes, as can be checked by differentiation,

$$\phi(t)=\phi_0(t)+\int_0^t \frac{d\tau}{\nu}\sin[m_1\nu_r(t-\tau)]\{(\nu^2-\omega_o^2)\phi + R\phi^3(\tau)+B\cos(m_2\nu_r\tau)\}, \tag{16}$$

where
$$\phi_0(t) = \phi_o \cos(m_1\nu_r t). \tag{17}$$

All terms obtained by iterating (16), starting with (17), are of the form $\sin(n\nu_r t)$ or $\cos(n\nu_r t)$. In the denominators we now have $(n\nu_r)$, with $n\neq 0$ [the case with n=0 in the integrand of (16) yields the secular terms discussed before and is eliminated by renormalizing the ν_r]. All denominators then have an (absolute) lower bound and the resulting series solutions of (11), and coupled systems in general, are rapidly convergent[1]. The driven oscillator of our example (11) is known as "The Duffing Equation" and has many peculiar properties, discussed in ref. 1. The solution of the two-coupled anharmonic oscillators (2) is developed here, in section V,

using the Carleman embedding discussed below.
Detailed quantitative results for a Hamiltonian system represented by (2) will be presented elsewhere[7] by C. Eminhizer and one of the authors. Finally, note that all solutions thus obtained have a (Poincaré) Recurrence Time $T_r=2\pi/\nu_r$, which may differ from the times $2\pi/\gamma$, $2\pi/\omega_o$ and $2\pi/\nu$ by any arbitrary number.

I.3 The Carleman Embedding

While the original Poincaré hierarchy constitutes some embedding of the nonlinear differential equation in an infinite set of linear differential equations, as we saw after eq. (9), it still requires one nonlinear operation per linear differential equation in order to find the inhomogeneous (driving) term. The same applies to any of our discrete recursion formulae, for the Fourier coefficients, in references 1, 6 and 7.

The main aim of this paper is to present a formalism similar to that of reference 1, i.e. avoiding secular terms and small denominators, but one in which the recursion formulae are *linear* in the Fourier coefficients. The formalism is based on certain ideas of Carleman, developed in 1931-32 in a paper[2] which was one of his last. Carleman's work was motivated by a remark of Poincaré's, at a 1908 conference in Rome, that one should be able to apply the theory of *linear* integral equations to the study of ordinary non linear differential equations [our eq. (3) is a *non* linear integral equation]. In this direction, Carleman proposed a scheme which embeds a single (or a system of) nonlinear differential equations in an *infinite* set of *linear* equations which, for their solution, no longer require nonlinear operations on intermediate results. Let us consider a very simple nonlinear rate equation as an example

$$\dot{\phi} = -\phi + \phi^2 \quad , \quad \text{with } \phi(0) = c, \tag{18}$$

and define new variables

$$y_1 \equiv \phi \ , \ y_2 \equiv \phi^2 \ ,\ldots, \ y_\lambda \equiv \phi^\lambda,\ldots, \tag{19}$$

where $\lambda = 1,2,3,\ldots$. Then

$$\dot{y}_\lambda = \lambda\phi^{\lambda-1}\dot{\phi} = \lambda(-\phi^\lambda + \phi^{\lambda+1}) = \lambda(-y_\lambda + y_{\lambda+1}), \tag{20}$$

where we used (18). Thus *eq. (18) has been embedded in an infinite set of equations (20) which are linear in the variables* y_λ . We rewrite (20) in vector form

$$\dot{\vec{y}} = M\vec{y} \quad , \quad \text{where} \tag{21}$$

$$M = \begin{pmatrix} -1 & 1 & 0 & 0 & \dots \\ 0 & -2 & 2 & 0 & \dots \\ 0 & 0 & -3 & 3 & \\ \dots & & & & \end{pmatrix} , \text{ and } (\vec{y})_\lambda \equiv y_\lambda , \qquad (22)$$

M is an infinite matrix. So our basic equations (1) and (2) can also be written in the linear form (21), by splitting a second order equation into two first order ones. This will be done in sections III, IV and V [The particular example (18) will be completely solved in section II]. Our periodic solutions of (1) and (2) correspond to the case where M has purely imaginary eigenvalues $\pm i\nu_r, \pm 2i\nu_r, \pm 3i\nu_r, \dots$ If we decompose $\vec{y}(t)$ into a discrete Fourier series eq. (21) becomes a time-independent *linear* vector equation for the Fourier coefficients. In order to solve this we must *first* find the eigenvalues of M which, in general, is a nonlinear problem itself. However, the problem of finding the Fourier frequencies has now been *separated* from that of finding the Fourier coefficients. The latter problem is easily solved, using (21), once the frequencies are found. Clearly the Carleman embedding scheme is extremely useful for those classes of nonlinear differential equations which yield an M whose eigenvalues are known or easily found, e.g. the example given in section II. While the perturbation theory in sections III-V is similar to that used in reference 1, work is in progress applying more direct matrix inversion methods, like the one of section II.

This formalism might also be better adapted to the matrix methods commonly used in quantum mechanics and other linear theories. The situation itself is also reminiscent of that in quantum mechanics where it is often easier to solve the eigenvalue problem for the, inherently, linear Schrödinger Equation than to solve the corresponding nonlinear Heisenberg (or even the classical) Equations. It should be mentioned that Carleman already investigated some of these questions using the, then, newly developed Hilbert Space Methods. Finally, we would like to point out that if, in addition to the eigenvalues of M, a complete set of eigenvectors of M is known we no longer have to restrict ourselves to the *periodic* solutions of nonlinear systems, as we do in section I.1. For in that case the solution of (21) can be immediately obtained and we find $\phi(t)$ as a series

$$\phi(t) \equiv y_1(t) = \sum_n A(n) \exp(i\mu_n t) \quad , \qquad (23)$$

with $i\mu_n$ the n-th eigenvalue of M. The μ_n need not be multiples of some ν_r as in section I.1, cf. (15) and (29) and we might obtain the "quasi periodic" solutions [i.e. when a finite number of the $\{\mu_n\}$ are rationally independent] as well; or even "almost-periodic" solutions [all μ_n independent] for some equations.

In section II we solve eq. (18) using the Carleman scheme and a direct matrix inversion of the linear eq. (21). Virtually all the ideas introduced in this article are explained or used in sections I and II.

Section III contains a discussion of the dynamics of the "quartic" oscillator characterized by eq. (1). Following the ideas proposed in reference 1 we here expand the frequency of oscillation ν about some suitable ν_0, rather than about the harmonic frequency ω_o as was done in (6). To this end we embed (1) in a broader class of equations

$$\ddot{\phi} + \nu_0^2\phi = \varepsilon\{(\nu_0^2-\omega_o^2)\phi + R\phi^3\} \quad , \text{cf. (16)}. \tag{24}$$

Eqs. (24) and (1) become equivalent for $\varepsilon=1$. The frequency ν_0 is chosen at a later stage so that the homogeneous part of (24), i.e. the new harmonic oscillator which results from (24) for $\varepsilon=0$, will have a solution which better mimics the actual solution of (1) than does the *previous* homogeneous part, i.e. R=0 in (1). The solution $\phi(t)$ and the frequency ν are now represented as a series in ε

$$\nu = \nu_0 + \varepsilon\nu_1 + \varepsilon^2\nu_2 + \ldots \quad , \tag{25}$$

The value of ν_0 *is chosen such that* ν_1 *vanishes.* The series which are thus obtained converge for all periodic solutions of (1), i.e. for all θ with $-\infty < \theta < +1$, cf. (26). It is shown that ν_1 disappears if

$$\nu_0^2 = \omega_o^2(1-\frac{3}{4}\theta) \quad , \quad \text{with} \quad \theta \equiv R\phi_o^2/\omega_o^2 \; , \tag{26}$$

for $p_o=0$ in (4), and the frequency of oscillation ν becomes

$$\nu = \nu_0\{1-\frac{3}{4}\varepsilon^2\xi_0^2 - \frac{69}{64}\varepsilon^4\xi_0^4 - \frac{633}{256}\varepsilon^6\xi_0^6 - \ldots\}, \qquad \text{cf. (89)}, \tag{27}$$

$$\text{with} \quad \xi_0 \equiv (\theta/8)/[1-\frac{3}{4}\theta] \tag{28}$$

The attractive feature of (27) is that the correction term to ν_0 is very small over a remarkably large range of initial conditions, i.e. $-\infty < \theta \lesssim + 0.8$. It is obvious from (27-28) that this is so for small values of $|\theta|$. However even as $\theta \to -\infty$ the first correction term never exceeds $(1/48)\varepsilon^2$ in absolute value, which is still small at $\varepsilon=1$. Higher order terms yield even better results.

In section IV a novel approach is taken to the solution of (1), following the ideas of reference 1. We note from (26-28) that ν is a function of ϕ_o and the parameters of eq. (24). One might imagine this function inverted so as to obtain ϕ_o as a function of ν and the parameters of (24). In that case, one has to *specify the value of* ν *in order to obtain the initial displacement* ϕ_o. In section IV we adjust our schemes so as to yield this inverted functional relation. We rewrite (24), using the prescribed ν, i.e. the actual frequency of oscillation, instead of the parameter ν_0. A unique solution $\phi(t)$ is obtained from which we recover $\phi_o=\phi(0)$. Hence the initial conditions are the derived quantities in this scheme. With this scheme we obtain all periodic solutions within one and the same region of convergence. Its real importance to us, however, lies in the fact that, for a *system* of oscillators, we can now also *specify* the actual frequencies. We found, in section I.2, that in order to avoid divergencies in our perturbation series for a system the ratio σ of any two frequencies must be rational. This would be impossible in the previous scheme (section III) where each higher order perturbation term adds a small correction to the frequency, cf. (25). It is only in this "inverted" scheme that we can specify the frequencies according to (13-15), unaltered by any corrections[1,7].

An application of this to our basic system in eq. (2) is given in section V. The displacements $\phi(\mu,t)$ can now be expanded in a *single-frequency* (Fourier) series

$$\phi(\mu,t) = \sum_{n=-\infty}^{+\infty} A_\mu(n)\exp(in\nu_r t), \quad \mu=1,2, \tag{29}$$

in accordance with eqs. (13-15).

The formalism developed in this paper will be applied to several specific problems in other reports. The approach given here differs from that of reference 1 in that the basic recursion formulae are linear and might therefore be investigated by techniques generally available in the theory of infinite matrices.

II. ON THE EMBEDDING OF NONLINEAR RATE EQUATIONS IN AN INFINITE SET OF LINEAR EQUATIONS

In this section we introduce an idea of T. Carleman for linearizing a nonlinear process, as usual, at the expense of replacing a *non*linear system described by a small number of variables with a *linear* one described by an *infinite* number of variables.

We present Carleman's general approach in its simplest form by, again, considering the elementary rate equation (18)

$$\dot{\phi} = -\phi + \phi^2, \text{ with } \phi(0) = c, \tag{30}$$

whose solution, as easily obtained by separating variables, is

$$\phi(t) = c[c+(1-c)e^t]^{-1}. \tag{31}$$

This solution can be obtained in a more complicated way by embedding (30) in a set of linear equations. While it is a ridiculous way to study equations which can be solved by elementary tricks, it may be a powerful general method to investigate equations which are *in*tractable using simple devices. Let

$$y_1 \equiv \phi,\ y_2 \equiv \phi^2, \ldots,\ y_\lambda \equiv \phi^\lambda, \ldots,\ \lambda = 1,2,3,\ldots \tag{32}$$

Then, according to (20), we obtain

$$\frac{dy_\lambda}{dt} = -\lambda y_\lambda + \lambda y_{\lambda+1}, \quad \text{with} \quad y_\lambda(0) = c^\lambda, \text{ or} \tag{33}$$

$$\dot{\vec{y}} = M\dot{\vec{y}}, \text{ with } (\vec{y})_\lambda \equiv y_\lambda, \text{ and } M, \tag{34}$$

the infinite matrix M of eq. (22).

One can then use his favorite method for solving this set of equations. The method of Laplace transforms will be used here. Now let

$$Y_\lambda(s) \equiv \int_0^\infty e^{-st}\, y_\lambda(t)dt. \tag{35}$$

Then, since

$$\int_0^\infty e^{-st}(dy_\lambda/dt)dt = -c^\lambda + sY_\lambda(s), \tag{36}$$

Eq. (33) becomes

$$(s+\lambda)Y_\lambda - \lambda Y_{\lambda+1} = c^\lambda, \tag{37}$$

or in vector form

$$(sI-M)\vec{Y} = \vec{c}, \text{ i.e.} \tag{38}$$

$$\begin{pmatrix} (s+1) & -1 & 0 & \cdots \\ 0 & (s+2) & -2 & \cdots \\ 0 & 0 & (s+3) & \cdots \\ \cdot & \cdot & \cdot & \cdots \\ \cdot & \cdot & \cdot & \cdots \end{pmatrix} \begin{pmatrix} Y_1 \\ Y_2 \\ Y_3 \\ \cdot \\ \cdot \end{pmatrix} = \begin{pmatrix} c \\ c^2 \\ c^3 \\ \cdot \\ \cdot \end{pmatrix}. \tag{39}$$

It is easy to construct the inverse of (38):

$$\vec{Y} = (sI-M)^{-1}\vec{c}, \quad \text{with} \tag{40}$$

$$(sI-M)^{-1} = \begin{pmatrix} \frac{1}{s+1} & \frac{1}{(s+1)(s+2)} & \frac{2\cdot 1}{(s+1)(s+2)(s+3)} & \cdot \\ 0 & \frac{1}{s+2} & \frac{2}{(s+2)(s+3)} & \cdot \\ 0 & 0 & \frac{1}{s+3} & \cdot \\ \cdot & \cdot & \cdot & \cdot \end{pmatrix}, \tag{41}$$

so that we have the following series solution, which is convergent when $|c/s|<1$,

$$Y_1 = \frac{c}{(1+s)} + \frac{c^2}{(2+s)(1+s)} + \frac{2!c^3}{(3+s)(2+s)(1+s)} + \ldots \tag{42}$$

$$= \sum_{k=1}^{\infty} c^k\, B(k,s+1), \tag{43}$$

$B(k,s+1)$ being the beta function defined by

$$B(k,j) \equiv \int_0^1 x^{k-1}(1-x)^{j-1}dx = \frac{\Gamma(k)\Gamma(j)}{\Gamma(k+j)} . \tag{44}$$

The series obtained by substituting (44) in (43), when $|c|<1$, sums to

$$Y_1 = \int_0^1 c(1-cx)^{-1}(1-x)^s dx. \tag{45}$$

Now let $(1-x) \equiv \exp(-t)$. Then

$$Y_1(s) = \int_0^{\infty} \frac{c\ e^{-st}dt}{c+(1-c)\exp\ t} \tag{46}$$

A comparison with the definition of $Y_1(s)$, eq. (35), yields the following identification, when $|c|<1$,

$$\phi(t) \equiv y_1(t) = \frac{c}{c+(1-c)e^t} \tag{47}$$

which is equivalent to (31) in the appropriate c range*

Notice that the matrix (sI-M) in (39) is a triangular matrix

---

*When $|c|>1$ convergent series are obtained by transforming to $z_\lambda \equiv \phi^{-\lambda}$, finding the Carleman set of linear equations for dz_λ/dt and treating the set in the manner that (33) was treated. One again obtains (47).

with no elements below the diagonal. If the right hand side of (30) were the polynomial

$$f(\phi) = -\phi + a_2\phi^2 + \ldots + a_L\phi^L \quad ,$$

the matrix (sI-M) would still be triangular but with L-1 filled diagonal lines above the central diagonal.

Equation (33) was derived from eq. (30) by Bellman and Richardson[8]; however, they did not discuss the construction of the solutions of the equations but used the equations to investigate the errors resulting from truncating the infinite set by making it finite. The truncation technique has of course been used in discussing many-body problems as well as hydrodynamic ones.

In the next section we generalize the Carleman embedding process to cover dynamical equations with *second* derivatives. However, in the remainder of this paper the resulting infinite set of linear equations will be solved by iterative techniques rather than by matrix inversion. Work is in progress applying similar *direct* matrix inversion methods to these and other dynamical equations.

III. DYNAMICS OF AN ANHARMONIC OSCILLATOR AND THE AVOIDANCE OF SECULAR TERMS

Let us consider the anharmonic oscillator, of unit mass, characterized by the Hamiltonian

$$H = \frac{1}{2}p^2 + \frac{1}{2}\omega_o^2\phi^2 - \frac{1}{4}R\phi^4 \quad , \tag{48}$$

with Hamilton's equations of motion

$$\dot{p} = \partial H/\partial\phi = -\omega_o^2\phi + R\phi^3 \quad \text{and} \quad \dot{\phi} = \partial H/\partial p = p \tag{49}$$

which are equivalent to

$$\ddot{\phi} + \omega_o^2\phi = R\phi^3 \tag{50}$$

The initial conditions are

$$x(0) = x_o \text{ and } p(0) = p_o \tag{51}$$

The Hamiltonian is conserved and has the value

$$\frac{1}{2}p_o^2 + \frac{1}{2}\omega_o^2\phi_o^2 - \frac{1}{4}R\phi_o^4$$

We will choose the origin of the time such that $p_o=0$ [6]. When $p_o\equiv0$, periodic solutions (50) to exist if

$$-\infty < \theta < +1, \quad \text{with} \quad \theta \equiv R\phi_o^2/\omega_o^2 \tag{52}$$

We now embed (52) in a broader class of equations

$$\ddot{\phi} + \nu_0^2\phi = \varepsilon\{(\nu_0^2-\omega_o^2)\phi + R\phi^3\} \tag{53}$$

cf. eqs. (24-28). For ε=1 eqs. (53) and (50) are identical. As explained in section I.3, we expand the frequency of oscillation about the same parameter ν_0 (so as to avoid secular terms)

$$\nu = \nu_0 + \varepsilon\nu_1 + \varepsilon^2\nu_2 + \dots , \tag{54}$$

cf. (25), instead of about the harmonic frequency ω_o, cf. (6). Using the language of reference 1, (53) will be called the *relocated* version of (50).

We now convert the *single* second order equation (53) into a *pair* of coupled first order nonlinear rate equations. These two equations will be embedded in an *infinite* set of *linear* equations, in the manner discussed in section I.3, cf.section II. Let us define a transformation

$$v \equiv \nu_0\phi + ip \quad \text{and} \quad w \equiv \nu_0\phi - ip = v^* \quad \text{or} \tag{55}$$

$$2\nu_0\phi = v + w \quad \text{and} \quad 2ip = v-w. \tag{56}$$

The relocated equation of motion (53) can then be written as

$$\dot{p} = -\nu_0^2\phi - \varepsilon\{S(v+w)-U(v+w)^3\} \quad \text{where} \tag{57}$$

$$S \equiv (\omega_o^2-\nu_0^2)/2\nu_0 \quad \text{and} \quad U \equiv R/8\nu_0^3 . \tag{58}$$

Since

$$\dot{v} = \nu_0\dot{\phi} + i\dot{p} \quad \text{and} \quad \dot{w} = \nu_0\dot{\phi} - i\dot{p} , \tag{59}$$

and using $\dot{\phi}$ = p we arrive at our pair of nonlinear rate equations

$$\dot{v} + i\nu_0 v = -i\varepsilon\{S(v+w) - U(v+w)^3\} \quad \text{and} \tag{60}$$

$$\dot{w} - i\nu_0 w = +i\varepsilon\{S(v+w) - U(v+w)^3\} , \tag{61}$$

noting that $w=v^*$.

Equations (60-61) are embedded in a set of linear equations by transforming to *new variables*

$$z_{\lambda\rho} \equiv v^{\lambda} w^{\rho} \quad , \quad \text{with} \quad \lambda,\rho = 0,1,2,3,\ldots, \tag{62}$$

analogous to the y_λ variables in (19) and (32). Hence with the aid of (60-61) we obtain new equations of motion

$$\dot{z}_{\lambda\rho} = \lambda v^{\lambda-1} w^{\rho} \dot{v} + \rho v\, v^{\lambda} w^{\rho-1} \dot{w}$$

$$\dot{z}_{\lambda\rho} - i\nu_0(\rho-\lambda) z_{\lambda\rho} = -i\varepsilon S[(\lambda-\rho) z_{\lambda\rho} + \lambda z_{\lambda-1,\rho+1} - \rho z_{\lambda+1,\rho-1}] +$$

$$i\varepsilon U[\lambda z_{\lambda-1,\rho+3} + (3\lambda-\rho) z_{\lambda,\rho+2} + (\lambda-3\rho) z_{\lambda+2,\rho}$$

$$- \rho z_{\lambda+3,\rho-1} + 3(\lambda-\rho) z_{\lambda+1,\rho+1}] \quad , \tag{63}$$

Thus eq. (53) is now embedded in an *infinite* set of equations (63) which are *linear* in the variables $z_{\lambda\rho}$. While we proceed as though we are going to find the solution of this infinite set of equations, our main interest is in the two variables $z_{1,0}$ and $z_{0,1}$ since their sum is proportional to $\phi(t)$ and their difference to $p(t)$.

According to the discussion in section I.1 we attempt to solve (63) by a Fourier series of frequency ν, i.e.

$$z_{\lambda\rho}(t) = \sum_{j=0}^{\infty} \varepsilon^{j} \sum_{n=-\infty}^{\infty} A_{\lambda\rho}(j,n)\eta^{n} \quad , \quad \text{with} \tag{64}$$

$$\eta \equiv \exp[i(\nu t+\delta)] \quad , \tag{65}$$

$$A^{*}_{\lambda\rho}(j,n) = A_{\rho\lambda}(j,-n) \quad , \quad \text{since } w=v^{*} \; ; \tag{66}$$

where the frequency ν and the phase shift δ depend on the initial conditions and ε. We substitute the series (64) in eq. (63) and equate like powers of ε and η. Using the series expansion (54) of $\nu(\varepsilon)$, we equate the coefficients of $\varepsilon^{j}\eta^{n}$ on both sides of eq. (63) and obtain our *basic, linear, formula*

$$(n-\rho+\lambda)\nu_0 A_{\lambda\rho}(j,n) + (1-\delta_{j,0})n\{\nu_1 A_{\lambda\rho}(j-1,n)+\ldots+\nu_j A_{\lambda\rho}(0,n)\} =$$

$$= -S[(\lambda-\rho)A_{\lambda\rho}(j-1,n)+\lambda A_{\lambda-1,\rho+1}(j-1,n)-\rho A_{\lambda+1,\rho-1}(j-1,n)]+$$

$$+U[\lambda A_{\lambda-1,\rho+3}(j-1,n)+(3\lambda-\rho)A_{\lambda,\rho+2}(j-1,n)+$$

$$3(\lambda-\rho)A_{\lambda+1,\rho+1}(j-1,n) + (\lambda-3\rho)A_{\lambda+2,\rho}(j-1,n)-\rho A_{\lambda+3,\rho-1}(j-1,n)] \quad (67)$$

If we keep the j-th order A and ν_j on the l.h.s., while bringing the remaining lower order terms over to the r.h.s. we obtain a *recursion formula,* i.e. a formula which calculates the j-th order coefficients from the, known, lower order coefficients on the r.h.s.

We start iterating (67) at j=0 and, in that case, find the first term of (67) to be the only nonvanishing term. Therefore, we can only satisfy eq. (67), nontrivially, if $n=\rho-\lambda$. Hence *the only nonvanishing zeroth order* A'*s are*

$$A_{\lambda\rho}(0,\rho-\lambda) \neq 0 \qquad ; \lambda,\rho = 0,1,2,3,\ldots, \quad (68)$$

This restriction can easily be removed, when necessary, by changing to the iteration scheme described in section 5.2 of reference 1 which does not employ power series in ε . The values of the coefficients (68) will be derived from the initial conditions in (73).

At j=1, we first consider the case $n=\rho-\lambda$. The first term in (67) is again zero, as are many terms on the r.h.s. due to (68), and we obtain

$$\nu_1 = S - 3U[A_{\lambda+1,\rho+1}(0,\rho-\lambda)/A_{\lambda\rho}(0,\rho-\lambda)] \quad , \quad (69)$$

our first order frequency correction. We show that this ratio of the A's is independent of λ or ρ as a result of the initial conditions.

III.1 Initial Conditions

At zeroth order, i.e. j=0, the initial conditions (51) are easily satisfied. Let us choose the phase angle δ, of (65), as

$$\tan(\delta) = -p_o/\nu_0\phi_o \quad \text{and} \quad f^2 \equiv \nu_0^2\phi_o^2 + p_o^2, \quad \text{whence} \quad (70)$$

$$v(0) = \nu_0\phi_o + ip_o = f\exp(-i\delta) \quad \text{and} \quad w(0) = v(0)^*. \quad (71)$$

The reason for this choice of phase angle becomes clear when we compare

$$z_{\lambda\rho}(0) = v(0)^\lambda w(0)^\rho = f^{\lambda+\rho} e^{i\delta(\rho \ \lambda)} \qquad \text{with}$$

$$z_{\lambda\rho}(0) = A_{\lambda\rho}(0,\rho-\lambda)e^{i\delta(\rho-\lambda)} \quad (72)$$

Therefore the zeroth order Fourier coefficients become

$$A_{\lambda\rho}(0,n) = \delta_{n,\rho-\lambda} f^{\lambda+\rho} , \tag{73}$$

cf. (68) and are real numbers. There is some freedom in the way we can satisfy the initial conditions at higher order [6]. We proceed here in a manner such that the initial conditions are exactly met at zeroth order, while each higher order is made to vanish at t=0, i.e. at j-th order

$$\sum_n A_{\lambda\rho}(j,n) \exp(in\delta) = 0 \qquad , \text{ for } j \geq 1. \tag{74}$$

This can always be achieved by *choosing* the value of $A_{\lambda\rho}(j,\rho-\lambda)$ so as to satisfy (74) [1]. Note that this coefficient was left arbitrary by the recursion relation (67).

It follows from (73) and (69) that ν_1 is independent of λ and ρ,

$$\nu_1 = S - 3\,Uf^2 \tag{75}$$

The quantities S, U and f are all functions of the parameter ν_0, cf. (58) and (70), which is still undetermined at this stage. As discussed before, cf. (25-26), *we choose* ν_0 *such that* ν_1 *vanishes* and: $\nu = \nu_0 + O(\varepsilon^2)$. Solving for ν_0^2 from (75), with $\nu_1=0$, we arrive at

$$\nu_0^2 = \tfrac{1}{2}\omega_o^2(1-\frac{3}{4}\theta) + \tfrac{1}{2}\{\omega_o^4(1-\frac{3}{4}\theta)^2-3Rp_o^2\}^{\frac{1}{2}} \quad , \text{ with} \tag{76}$$

$$\theta \equiv R\phi_o^2/\omega_o^2 \quad , \tag{77}$$

This reduces to

$$\nu_0^2 = \omega_o^2\,(1-\frac{3}{4}\theta) \quad , \text{ when } p_o = 0 . \tag{78}$$

As a result of this choice of ν_0 the solution of the unperturbed, homogeneous part of (53) is already "close" to the exact solution of (53) and we obtain convergence for the periodic solutions, cf.(89), i.e. $-\infty < \theta < +1$ [The unbounded solutions for $\theta > +1$ can be obtained in a similar way, see section 3.5 of reference 1].

III.2 Higher Order Contributions

Generally, if $S\neq 0$ (i.e. if $\nu_0\neq\omega_o$) the recursion formula (67) can be put in a form which does not contain any parameters by making the transformations

$$A_{\lambda\rho}(j,n)e^{in\delta} \equiv C_{\lambda\rho}(j,n)(U/\nu_0)^j(S/3U)^{\frac{1}{2}(\lambda+\rho+2j)}, \tag{79}$$

$$\nu_j \equiv \Omega_j\nu_0(S/3\nu_0)^j \quad , \tag{80}$$

with "initial conditions"

$$C_{\lambda\rho}(0,n) = \delta_{n,\rho-\lambda}\, e^{in\delta}(3Uf^2/S)^{\frac{1}{2}(\lambda+\rho)} \quad = \quad \delta_{n,\rho-\lambda}e^{in\delta} \; , \tag{81}$$

cf. (79) and (73). The latter equality applies due to our choice of $\nu_1=0$ in (75). The (S/3U) of (79) may then be replaced by the initial amplitude f^2. When $\nu_1=0$ we also have $\Omega_1=0$ and the linear recursion relation for the A's, (67), can be transformed into

$$(n-\rho+\lambda)C_{\lambda\rho}(j,n) + n\sum_{k=2}^{j}\Omega_k C_{\lambda\rho}(j-k,n) \quad =$$

$$-3[(\lambda-\rho)C_{\lambda\rho}(j-1,n)+\lambda C_{\lambda-1,\rho+1}(j-1,n)-\rho C_{\lambda+1,\rho-1}(j-1,n)] \; +$$

$$+\;\lambda C_{\lambda-1,\rho+3}(j-1,n)+(3\lambda-\rho)C_{\lambda,\rho+2}(j-1,n)+3(\lambda-\rho)C_{\lambda+1,\rho+1}(j-1,n) \; +$$

$$+\;(\lambda-3\rho)C_{\lambda+2,\rho}(j-1,n)-\rho C_{\lambda+3,\rho-1}(j-1,n) \tag{82}$$

The $C_{\lambda,\rho}(j,\rho-\lambda)$ is now solved from the transformed version of (74)

$$\sum_n C_{\lambda\rho}(j,n) = 0 \; , \tag{83}$$

once we obtain the other j-th order C's from (82). It is clear from (81) and (76) that the results become much simpler if $\delta=0$, i.e. $p_o=0$ in (70). As explained before (52) this is possible for any solution of (50).

In this case, $p_o=0$, we obtain the first order coefficients from eqs. (81) and (82) at j=1,

$$C_{\lambda\rho}(1,\rho-\lambda+k) = \tfrac{1}{4}\{\lambda\Lambda_{k,4}+\rho\Lambda_{k,-4}-2\rho\Lambda_{k,2}-2\lambda\Lambda_{k,\,2}\}, \tag{84}$$

$$\text{with } \Delta_{k,m} \equiv \delta_{k,m} - \delta_{k,0} = \begin{cases} +1 & \text{if } k=m\neq 0 \\ -1 & \text{if } k=0,\ m\neq 0 \\ 0 & \text{otherwise} \end{cases} \qquad (85)$$

where the k=0 result is desired with the aid of (83).

At j=2 and $n=\rho-\lambda$ we obtain the second order frequency correction from (82)

$$\Omega_2 = -3/4 \quad , \qquad (86)$$

where we used (81) and (84) [when $\delta\neq 0$, eq. (84) is slightly changed and the r.h.s. of (86) is $(3/4)(2\cos 4\delta - 4\cos 2\delta + 1)$]. At j=2 and $n\neq\rho-\lambda$ we obtain from (82) and (84)

$$\begin{aligned} 32C_{\lambda\rho}(2,\rho-\lambda+k) = {} & \lambda(\lambda-1)\delta_{k,8} + \rho(\rho-1)\delta_{k,-8} - 4\lambda(\rho+1)\delta_{k,6} - 4\rho(\lambda+1)\delta_{k,-6} + \\ & + 2(\rho-\lambda+\lambda^2+2\rho^2+\rho\lambda)\delta_{k,4} + 2(\lambda-\rho+\rho^2+2\lambda^2+\lambda\rho)\delta_{k,-4} + \\ & + 4(\rho+4\lambda-\lambda^2-\rho^2-\rho\lambda)\delta_{k,2} + 4(\lambda+4\rho-\rho^2-\lambda^2-\lambda\rho)\delta_{k,-2} \end{aligned} \qquad (87)$$

Using (83) we then also find

$$32C_{\lambda\rho}(2,\rho-\lambda) = \lambda^2+\rho^2-15(\lambda+\rho) + 12\rho\lambda \qquad (88)$$

Higher order Ω's and C's have been obtained from the recursion relations (82) by a computer calculation. The program has kindly been prepared for the authors by Ralph Janda. The frequency ν, to eighth order, is

$$\nu = \nu_0\{1 - \tfrac{3}{4}\varepsilon^2\xi_0^2 - \tfrac{69}{64}\varepsilon^4\xi_0^4 - \tfrac{633}{256}\varepsilon^6\xi_0^6 - \tfrac{110421}{16384}\varepsilon^8\xi_0^8 \ldots\}, \text{ with} \qquad (89)$$

$$\xi_0 \equiv S/3\nu_0 = (\omega_o^2-\nu_0^2)^2/6\nu_0^2 = \qquad (90)$$

$$= \tfrac{\theta}{8}/(1-\tfrac{3}{4}\theta) \quad , \quad \text{with} \quad \theta \equiv R\phi_o^2/\omega_o^2 \; , \qquad (91)$$

all for $p_o=0$. We consider the case $\varepsilon=1$, when (53) is equivalent to (50), and observe that the series (90) in ξ_0 is rapidly convergent for all values of θ with $-\infty < \theta < +1$, except for those close to the point of transition to unbounded solutions, $\theta = +1$. As $\theta \to -\infty$, $\xi_0 \to -\frac{1}{6}$ and (89) becomes

$$\nu = \nu_0\{1 - (1/48) - (23/27648) + \ldots\}, \text{ as } \theta \to -\infty, \quad \varepsilon=1, \tag{92}$$

where $\nu_0/\omega_o = (1-3\theta/4)^{1/2}$. For example, when $\theta = -10^3$, which is far from being a small anharmonicity, the first three terms of (89) yield

$$(\nu/\omega_o)^2 = 718.899\ldots , \tag{93}$$

while the five terms in (89) agree to six places with the exact value [1], 718.81569... The zeroth order approximation is $(\nu_0/\omega_o)^2 = 751.0$, already within a few percent of that value.

The Carleman linearization of (53) generates of course the same series in ξ_0, or θ, as obtained in (section 2 of) ref. 1 via a *non*linear recursion relation. Detailed numerical results are tabulated there.

The actual oscillator displacement $\phi(t)$ is obtained from $z_{1,0}$ and $z_{0,1}$, cf. (62) and (56), as

$$\phi(t)=(1/2\nu_0)\{z_{1,0}+z_{0,1}\}=(1/2\nu_0)\{\sum_j \varepsilon^j \sum_n [A_{1,0}(j,n)+A_{0,1}(j,n)]e^{in\nu t}\}, \tag{94}$$

for $p_o=0$, according to (64). These A coefficients were transformed into C coefficients in (79). The lower order C's are obtained from (81), (84) and (87-88)

$$C_{1,0}(0,n) = \delta_{n,-1} \quad , \tag{95}$$

$$4\, C_{1,0}(1,n) = \delta_{n,3} + \delta_{n,1} - 2\delta_{n,-3} \quad , \tag{96}$$

$$32\, C_{1,0}(2,n) = -4\delta_{n,5} + 12\delta_{n,1} - 14\delta_{n,-1} + 6\delta_{n,-5} \tag{97}$$

while the corresponding $C_{0,1}$'s are easily found from the above, using the symmetry relation (66), with (79). Higher order terms can be calculated from (95-97) through the recursion relations (82-83) and are listed in Table I below. From (94-97) and (79) we finally obtain, to fourth order, the displacement

$$\phi(t)=\phi_o\{c_1+(\xi_0/4)(c_1-c_3)+(\xi_0/4)^2(c_5-c_1)+(\xi_0/4)^3(19c_1-18c_3-c_7) + +(\xi_0/4)^2(c_9+38c_5-c_3-38c_1)+\ldots\}, \tag{98}$$

$$\text{with } c_n = \cos n\nu t \quad , \tag{99}$$

and the ξ_0 defined in (90-91). We can reorder (98) into the Fourier cosine series

$$\phi(t)/\phi_o = \sum_{j=0}^{\infty} c_{2j+1} \sum_{k=j}^{\infty} a_{jk} (\xi_0/4)^k , \tag{100}$$

where the coefficients a_{jk}, easily obtained from Table I and (66), are exhibited in Table II to eighth order in j.

n \ j	1	2	3	4	5	6	7	8
-1	1	-7	13	-170	428	-6418	17792	-285055
1	0	6	6	132	252	4698	11766	200214
-3	-2	0	-18	-2	-504	-138	-18664	-8412
3	1	0	0	1	-117	60	-7579	3297
-5		3	0	84	0	3327	3	151056
5		-2	0	-46	0	-1608	-2	-66334
-7			-4	0	-190	0	-9466	0
7			3	0	132	0	6249	0
-9				5	0	336	0	20121
9				-4	0	-258	0	-15006
-11					-6	0	-522	0
11					5	0	424	0
-13						7	0	748
13						-6	0	-630
-15							-8	0
15							7	0
-17								9
17								-8

Table I. Table of *integer* values of $4^j C_{1,0}(j,n)$ for $j=1,2,\ldots,8$ and $-17 \leq n \leq 17$, calculated with the aid of an electronic computer, using integer variables. Only odd values of n are excited and the C's vanish for $|n| > 2j+1$. The values of $C_{0,1}(j,n)$ are obtained from the table and the symmetry relation (66) with (79).

k \ j	1	2	3	4	5	6	7	8
0	1	-1	19	-38	680	-1720	29558	-84841
1	-1	0	-18	-1	-621	-78	-26243	-5115
2		1	0	38	0	1719	1	84722
3			-1	0	-58	0	-3217	0
4				1	0	78	0	5115
5					-1	0	-98	0
6						1	0	118
7							-1	0
8								1

Table II. The coefficients a_{jk} which appear in eq. (100).

IV. "BACKWARD SCHEME" ANALYSIS OF THE ANHARMONIC OSCILLATOR

In the previous section we derived the renormalized frequency ν of the anharmonic oscillator (1), as a function of the initial displacement ϕ_o (and the other parameters), cf. (89-91), so as to avoid the secular terms discussed in section I.1. Here we take a novel approach to the solution of (1) and derive the inverse of this functional relation, i.e. we *specify* ν and derive ϕ_o. Following the language of ref. 1 we will call this a *"Backward Scheme"*. Hence the initial conditions are *derived* quantities, in the backward scheme, which are recovered by setting t=0 in the solutions $\phi(t)$ and $\dot{\phi}(t)$. Many other schemes are possible, e.g. we could renormalize the parameter R in the Hamiltonian (48) *specifying* both ν and ϕ_o, etc. C. Ginsburg and one of the authors have made use of a similar renormalization to obtain the energy levels of some (polynomial) anharmonic quantum oscillators [9].

Since in a backward scheme the exact frequency of oscillation ν is specified we use the ν as the relocation parameter instead of some ν_0, as in (53), and arrive at

$$\ddot{\phi} + \nu^2\phi = \varepsilon\{(\nu^2-\omega_o^2)\phi + R\phi^3\} \quad , \tag{101}$$

analogous to (53) in the previous section. For $\varepsilon=1$ the relocated eq. (101) is identical to the original anharmonic oscillator (1). We then define

$$v \equiv \nu\phi + ip \quad \text{and} \quad w \equiv \nu\phi - ip = v^* \tag{102}$$

The two resulting rate equations are analogous to (60-61)

$$\dot{v} + i\nu v = -i\varepsilon\{S(v+w)-U(v+w)^3\} \quad \text{and} \quad \dot{w} = \dot{v}^* \quad , \tag{103}$$

$$S \equiv (\omega_o^2-\nu^2)/2\nu \quad \text{and} \quad U \equiv R/8\nu^3 \quad , \tag{104}$$

cf. (58). Again we obtain an infinite set of linear equations of motion for the $z_{\lambda\rho}$

$$\dot{z}_{\lambda\rho} - i\nu(\rho-\lambda)z_{\lambda\rho} = \text{"r.h.s.(63)"} \quad , \quad \text{with} \tag{105}$$

$$z_{\lambda\rho} \equiv v^\lambda w^\rho \quad . \qquad \lambda,\rho=0,1,2,\ldots \tag{106}$$

Since many formulae have *"the same functional form"* as the ones in the previous section we will not repeat them explicitly. Notice however that the old ν_0 must be replaced everywhere by ν [The previous v, w, S and U have been redefined explicitly in (102-104)].

Again we expand the z in a power series in ε and a Fourier series in ν, cf. (64-66). This time however the ν is specified and not expanded in a power series in ε. Hence the *basic, linear, recursion relation* for the coefficients $A_{\lambda\rho}(j,n)$ changes into

$$(n-\rho+\lambda)\nu A_{\lambda\rho}(j,n) = \text{"r.h.s.(67)"} \quad , \tag{107}$$

so the $A_{\lambda\rho}(0,\rho-\lambda)$ is again the only nonvanishing zeroth order coefficient.

At j=1, we first consider the "resonance condition" [1], i.e. $n=\rho-\lambda$ in (107), and obtain this time

$$0 = S - 3\,U[A_{\lambda+1,\rho+1}(0,\rho-\lambda)/A_{\lambda\rho}(0,\rho-\lambda)] \quad , \tag{108}$$

cf. (69)! We must satisfy (108) for all λ and ρ. Therefore we let

$$A_{\lambda\rho}(0,n) \equiv \delta_{n,\rho-\lambda} f^{\lambda+\rho} \quad \text{, whence} \tag{109}$$

$$f^2 = S/3U = 4\nu^2(\omega_o^2-\nu^2)/3R \quad , \tag{110}$$

due to (108). A parameterless recursion formula is obtained when we transform to new coefficients C with

$$A_{\lambda\rho}(j,n)e^{in\delta} \equiv C_{\lambda\rho}(j,n)(U/\nu)^j f^{(\lambda+\rho+2j)} \quad , \tag{111}$$

the analogue of (79),[when we chose $\nu_1=0$ there]. From (109) we find the "initial" C

$$C_{\lambda\rho}(0,n) = \delta_{n,\rho-\lambda} e^{in\delta} \tag{112}$$

The resulting recursion relation for the C's becomes

$$(n-\rho+\lambda)C_{\lambda\rho}(j,n) = \text{"r.h.s.(82)"} \quad , \tag{113}$$

which for j=0 and j=1 is identical to the previous eq. (82). The resonance condition, $n=\rho-\lambda$ in (113), at j=1, is automatically satisfied, owing to (108-112). Since the initial C's are the same also, cf. (112) and (81), we obtain the same first order coefficients, <u>for $k\neq 0$</u> in

$$C_{\lambda\rho}(1,\rho-\lambda+k) = \frac{1}{4}\{\lambda\delta_{k,4}+\rho\delta_{k,-4}-2\rho\delta_{k,2}-2\lambda\delta_{k,-2}\} \quad , \tag{114}$$

cf. (84-85). We have again restricted ourselves to the case with zero phase angle δ in (112), which *results* in $\phi(0)=0$ as can be seen from the Fourier series (64-66). However for k=0 we get a first order coefficient which is different from the one in the previous section (84-85), since we do *not* satisfy an initial condition $\phi(0)=\phi_o$ in the backward scheme. The special coefficients $C_{\lambda\rho}(1,\rho-\lambda)$ are not determined at j=1 by the recursion relation (113) but are found at j=2. At j=2 and $n=\rho-\lambda$ the l.h.s. of (113) vanishes. The r.h.s. contains the known first order terms (114) and the unknown special coefficients. Solving for these we find

$$C_{\lambda+1,\rho+1}(1,\rho-\lambda)-C_{\lambda\rho}(1,\rho-\lambda) = 1/4 \quad , \tag{115}$$

for all λ and ρ. Therefore the (λ,ρ) coefficient must be of the form $a\lambda+b\rho$. As a result of the symmetry relation (66) we have b=a and obtain from (115)

$$C_{\lambda\rho}(1,\rho-\lambda) = (\rho+\lambda)/8 \quad , \tag{116}$$

which is exactly *half* the value of the same coefficient (84-85) in the "forward" scheme of the previous section.

Therefore the second order coefficients, derived from (116), will be different also, from the ones (87-88) in the previous section. For $k\neq 0$, we now get

$$\begin{aligned}32C_{\lambda\rho}(2,\rho-\lambda+k) = {} & \lambda(\lambda-1)\delta_{k,8}+\rho(\rho-1)\delta_{k,-8}-4\lambda(\rho+1)\delta_{k,6}-4\rho(\lambda+1)\delta_{k,-6} + \\ & +(2\rho-4\lambda+\lambda^2+4\rho^2+\rho\lambda)\delta_{k,4}+(2\lambda-4\rho+\rho^2+4\lambda^2+\lambda\rho)\delta_{k,-4} + \\ & +2(4\rho+2\lambda-2\lambda^2-\rho^2-\rho\lambda)\delta_{k,2}+2(4\lambda+2\rho-2\rho^2-\lambda^2-\lambda\rho)\delta_{k,-2}\end{aligned} \tag{117}$$

The second order special coefficients are found from the recursion relation (113) at j=3 and $n=\rho-\lambda$. The l.h.s. vanishes again and solving for these coefficients we obtain

$$32[C_{\lambda+1,\rho+1}(2,\rho-\lambda)-C_{\lambda,\rho}(2,\rho-\lambda)] = 11(\rho+\lambda) + 4 \tag{118}$$

These C's must be quadratic in λ and ρ and have to satisfy the symmetry relation (66). If we postulate them to be of the form

$$C_{\lambda,\rho}(2,\rho-\lambda) = a(\rho+\lambda)+b(\rho+\lambda)^2 \quad , \tag{119}$$

we obtain, from (118)

$$C_{\lambda,\rho}(2,\rho-\lambda) = -7(\rho+\lambda)/64 + 11(\rho+\lambda)^2/128 \tag{120}$$

Using (112-120) we can calculate $\phi(t)$ to second order in ε. From (106) and (102) we obtain

$$\phi(t) = (z_{1,0} + z_{0,1})/2\nu \tag{121}$$

The z's were expanded in a Fourier series with coefficients $A_{\lambda\rho}$ or $C_{\lambda\rho}$, related by (111). So we only have to calculate the $C_{1,0}$'s [the $C_{0,1}$'s there follow from the symmetry relation (66)]. We have from (112-120):

$$C_{1,0}(0,n) = \delta_{n,\rho-\lambda} \tag{122}$$

$$4C_{1,0}(1,n) = \delta_{n,3} + \tfrac{1}{2}\,\delta_{n,1} - 2\delta_{n,-3} \tag{123}$$

$$32C_{1,0}(2,n) = 6\delta_{n,5} + 6\delta_{n,3} - \frac{3}{4}\,\delta_{n,1} - 3\delta_{n,-3} - 4\delta_{n,-5}\,, \tag{124}$$

cf. (95-97). Substituting these back into (111), (64) and (121), we arrive at

$$\phi(t)=(2a)\{c_0+(\xi/8)(c_1-2c_3)+(\xi^2/128)(8c_5+12c_3-3c_1)+\dots\}\,, \tag{125}$$

cf. (98-99), where we used ε=1 and defined

$$c_n \equiv \cos n(\nu t+\delta)\quad , \tag{126}$$

$$\xi \equiv S/3\nu = Uf^2/\nu = (\omega_o^2-\nu^2)/6\nu^2 \quad \text{and} \tag{127}$$

$$2a \equiv f/\nu \quad ; \quad \text{whence } \nu^2=\omega_o^2\,[1-\tfrac{3R}{4}(2a)^2/\omega_o^2], \tag{128}$$

cf. (77-78). If R>0 we must clearly *choose* $\nu < \omega_o$, according to (110) and (128), and $\nu > \omega_o$ if $R < 0$. When $R < 0$ the restoring force is stronger than the harmonic force alone, cf. (50). Hence we do expect the period to be shorter, i.e. $\nu > \omega_o$.

The expression (125) for $\phi(t)$ is a power series in ξ. The parameter ξ has an absolute value less than 1/6 for all ν with $\omega_o^2/2 < \nu^2 < \infty$, so that *the series* (125) *converges rapidly for a very large range of* ν *values*. In terms of the parameter θ of the previous section, cf. (77-78), this range is $-\infty < \theta \lesssim 2/3$, and the series converges rapidly for all periodic solutions except for those near $\theta = +1$, the point of transition to unbounded solutions.

Having obtained the $\phi(t)$ in (125) we are finally in a position to recover the "initial" conditions

$$\phi(0)=(2a)\{c_1+(\xi/8)(c_1-2c_3)+(\xi^2/128)(8c_5+12c_3-3c_1) + \ldots\} , \quad (129)$$

$$\dot\phi(0)=(2a\nu)\{-s_1+(\xi/8)(6s_3-s_1)+(\xi^2/128)(3s_1-36s_3-40s_1) + \ldots\}, \quad (130)$$

with $c_n \equiv \cos n\delta$ and $s_n \equiv \sin n\delta$ (131)

So in this backward scheme we *specify* the values of ν and δ, calculate ξ and a from (127-128), and obtain the above $\phi(0)$ and $\dot\phi(0)$ as the results. Consider the case with $\delta=0$, i.e. $p_o=0$. Then (129) yields

$$\phi(0) = (2a)\{1-\xi/8 + 17\xi^2/128 + \ldots\} \quad (132)$$

In the range $\frac{1}{2}\omega_o^2 < \nu^2 < \infty$ we noted $|\xi| < 1/6$. Hence, in this range, we have, to within a few percent

$$\phi(0) \approx 2a \quad , \quad \text{or} \quad (133)$$

$$\nu^2 \approx \omega_o^2(1-3\theta/4), \qquad \text{with } \theta \equiv R\phi^2(0)/\omega_o^2 , \quad (134)$$

the same approximation as obtained from the "forward" scheme of the previous section, cf. (89) and (78). We can even use (132) in a "forward" manner by specifying $\phi(0)$, solving for ξ and from ξ, (127), obtain a very close estimate of the frequency ν.

In this version of the backward scheme the calculation of the special coefficients $C_{\lambda\rho}(j,\rho-\lambda)$, cf. (115-120), is somewhat more involved than the corresponding calculation in the, previous, "forward" scheme, i.e. (84) and (88). The same applies to some extent to the versions described in section 3 of ref. 1. They can be placed on an equal footing by transforming the equation of motion (101) to

$$\ddot\chi + \nu^2\chi = \varepsilon[(\nu^2-\omega_o^2)\chi + \theta\chi^3] \qquad \text{with} \quad (135)$$

$$\chi = \phi(t)/\phi(0) \quad ; \quad \text{whence} \quad \chi(0) = 1 \quad \text{and} \quad (136)$$

$$\theta = \theta_0 + \varepsilon\theta_1 + \varepsilon^2\theta_2 + \ldots \quad , \quad (137)$$

i.e. at $n=\rho-\lambda$ and j-th order we now solve for θ_{j-1}. The special coefficients are obtained from the condition $\chi(0)=1$, (136), exactly like we did in the forward scheme, cf. (73-74). This procedure is used in section 5 and Table III of ref. 1 for our *non*linear recursion relations

V. AVOIDANCE OF SMALL DENOMINATORS IN COUPLED OSCILLATOR SYSTEMS

As was discussed in section I.2 of the introduction, a classical difficulty in the traditional perturbation theory of coupled anharmonic oscillator *systems* is the appearance of "Small Denominators" which prevent the convergence of the perturbation series. We avoid this problem by solving the system for such initial conditions that the frequency of oscillation of *each* oscillator is an integer multiple of some *recurrence frequency* ν_r, cf. (15). Clearly the *backward* scheme, in which we *specify* these frequencies, is most suited to this end.

Here we show how this scheme can be combined with the Carleman embedding procedure to yield linear recursion formulae for the Fourier coefficients.

Let us consider the two coupled anharmonic oscillators characterized by eq. (2). We specify the two main frequencies of oscillation, ν_1 and ν_2, as

$$\nu_1 \equiv m_1\nu_r \quad \text{and } \nu_2 \equiv m_2\nu_r \ , \tag{138}$$

$$\text{with } \ \sigma \equiv \nu_2/\nu_1 = m_2/m_1 \ , \tag{139}$$

a rational number. The values of ν_1, ν_2 and phase angles δ_1, δ_2 will be the "initial" conditions for this backward scheme. The relocated version of (2), analogous to (101), is

$$\ddot{\phi}(\mu)+\nu_\mu^2\phi(\mu)=\varepsilon\{[\nu_\mu^2-\omega_o^2(\mu)]\phi(\mu)+R\sum_{k=0}^{3} a_{\mu k}[\phi(1)]^{3-k}[\phi(2)]^k\}, \tag{140}$$

with $\mu=1,2$. These can now be converted into four coupled rate equations by defining, for $\mu=1,2$,

$$v_\mu \equiv \nu_\mu\phi(\mu) + ip_\mu \quad \text{and} \quad w_\mu \equiv v_\mu^* \ , \tag{141}$$

with $p_\mu \equiv \dot{\phi}(\mu)$, cf. (102). Using the definition

$$S_\mu \equiv [\omega_o^2(\mu) - \nu_\mu^2]/2\nu_\mu \ , \tag{142}$$

cf. (104), and new coefficients $U_\mu(\alpha_1,\beta_1,\alpha_2,\beta_2)$, to be defined in

Table III below, we obtain the rate equations

$$\dot{v}_\mu + i\nu_\mu v_\mu = -i\varepsilon\{S_\mu(v_\mu+w_\mu) - \sum_{(\alpha;\beta)} U_\mu(\alpha_1,\beta_1,\alpha_2,\beta_2)v_1^{\alpha_1}w_1^{\beta_1}v_2^{\alpha_2}w_2^{\beta_2}\} \quad (143)$$

$$\text{and} \quad \dot{w}_\mu = \dot{v}_\mu^* \quad , \quad \text{for } \mu = 1,2, \quad (144)$$

cf. (103). The $(\alpha;\beta)$ summation extends over all sets $(\alpha_1,\beta_1,\alpha_2,\beta_2)$ such that

$$\alpha_1 + \beta_1 + \alpha_2 + \beta_2 = 3 \quad , \quad \text{while } 0 \leq \{\alpha_\mu,\beta_\mu\} \leq 3 \quad (145)$$

The explicit values of $U_\mu(\alpha;\beta)$ are given in Table III.

α_1	β_1	α_2	β_2	U_1	$\beta_2-\alpha_2$	$1+\beta_1-\alpha_1$	α_1	β_1	α_2	β_2	U_1	$\beta_2-\alpha_2$	$\beta_1-\alpha_1+1$
3	0	0	0	b_{10}	0	-2	1	0	2	0	b_{12}	-2	0
0	3	0	0	b_{10}	0	4	0	1	2	0	b_{12}	-2	2
0	0	3	0	b_{13}	-3	1	0	0	2	1	$3b_{13}$	-1	1
0	0	0	3	b_{13}	3	1	1	0	0	2	b_{12}	2	0
2	1	0	0	$3b_{10}$	0	0	0	1	0	2	b_{12}	2	2
2	0	1	0	b_{11}	-1	-1	0	0	1	2	$3b_{13}$	1	1
2	0	0	1	b_{11}	1	-1	1	1	1	0	$2b_{11}$	-1	1
1	2	0	0	$3b_{10}$	0	2	1	1	0	1	$2b_{11}$	1	1
0	2	1	0	b_{11}	-1	3	1	0	1	1	$2b_{12}$	0	0
0	2	0	1	b_{11}	1	3	0	1	1	1	$2b_{12}$	0	2

Table III. Table of the coefficients $U_1(\alpha_1,\beta_1,\alpha_2,\beta_2)$ which appear in equations (143) and (148) of section V. The $b_{\mu k}$ in the table are related to the $a_{\mu k}$ of eq. (140) by $b_{\mu k} \equiv Ra_{\mu k}/(2\nu_1)^{3-k}(2\nu_2)^k$. The coefficients $U_2(\alpha_1,\beta_1,\alpha_2,\beta_2)$ are derivable from U_1's of the Table by replacing $b_{1,k}$ by $b_{2,k}$.

The U's satisfy the symmetry relation

$$U_\mu(\alpha_1,\beta_1,\alpha_2,\beta_2) = U_\mu(\beta_1,\alpha_1,\alpha_2,\beta_2) = U_\mu(\alpha_1,\beta_1,\beta_2,\alpha_2) \quad (146)$$

Our four rate equations (143-144) can be embedded in an infinite set of *linear* rate equations in the usual way, cf. (105-106), by transforming to

$$z_{\lambda_1,\rho_1,\lambda_2,\rho_2} \equiv v_1^{\lambda_1} w_1^{\rho_1} v_2^{\lambda_2} w_2^{\rho_2} \quad , \tag{147}$$

denoted symbolically by $z_{\lambda;\rho}$. Eqs. (143-144) become

$$\left[\frac{d}{dt} - i\,\nu\cdot(\rho-\lambda)\right] z_{\lambda;\rho} = -i\varepsilon \sum_{\mu=1}^{2} \Big[S_\mu[(\lambda_\mu-\rho_\mu)+\lambda_\mu E_\mu^{-1}F_\mu - \rho_\mu E_\mu F_\mu^{-1}] + - \sum_{(\alpha;\beta)} U_\mu(\alpha;\beta)[\lambda_\mu E_\mu^{-1} - \rho_\mu F_\mu^{-1}] E_1^{\alpha_1} F_1^{\beta_1} E_2^{\alpha_2} F_2^{\beta_2} \Big] z_{\lambda;\rho} \quad , \tag{148}$$

cf. (105) and (63), where we used the index raising and lowering operators E and F with

$$E_1^{\pm1} z_{\lambda_1,\rho_1,\lambda_2,\rho_2} \equiv z_{\lambda_1\pm1,\rho_1,\lambda_2,\rho_2} \quad \text{and} \tag{149}$$

$$F_1^{\pm1} z_{\lambda_1,\rho_1,\lambda_2,\rho_2} \equiv z_{\lambda_1,\rho_1\pm1,\lambda_2,\rho_2} \quad , \text{ etc.,} \tag{150}$$

and where we introduced a vector notation

$$\rho\cdot\nu \equiv \rho_1\nu_1 + \rho_2\nu_2 = \rho\cdot m\,\nu_r \quad , \tag{151}$$

because of (138). Since both frequencies are multiples of ν_r, cf. (138), we can, as discussed in section I.2 of the introduction, expand the z's in a discrete Fourier series of frequency ν_r

$$z_{\lambda;\rho} = \sum_{j=0}^{\infty} \varepsilon^j \sum_{n=-\infty}^{\infty} A_{\lambda;\rho}(j;n)\,\eta^n \quad , \text{ with} \tag{152}$$

$$\eta \equiv \exp[i(\nu_r t + \delta_r)] \quad , \text{ and} \tag{153}$$

$$A^*_{\lambda;\rho}(j,n) = A_{\rho;\lambda}(j,-n) \quad , \text{ since } w_\mu = v_\mu^* \quad , \tag{154}$$

cf. (64-66) and (29). As an example of the magnitudes of ν_1,ν_2, note that ν_r=1 when (ν_1,ν_2)=(1,3) while $\nu_r=10^{+20}$ when (ν_1,ν_2)= $(1+10^{-20},3)$, i.e. *small changes in* ν_1 *and* ν_2 *can result in tremendous changes in* ν_r. By substituting (152) in the equation of motion

(148) and equating the coefficients of $\varepsilon^j\eta^n$ on both sides we can again obtain a basic, linear, recursion formula for the A coefficients, cf. (107) and (67). As in the previous sections we find it more convenient to first transform to C coefficients defined by

$$A_{\lambda;\rho}(j;n)\zeta_\mu^{\lambda_\mu-\rho_\mu} \equiv C_{\lambda;\rho}(j;n) \prod_{\mu=1}^{2} f_\mu^{\lambda_\mu+\rho_\mu+2j} \quad , \quad \text{with} \tag{155}$$

$$f_1^2 = (3S_1b_{23}-2S_2b_{12})/D \quad \text{and} \quad f_2^2 = (3S_2b_{10}-2S_1b_{21})/D, \tag{156}$$

$$\zeta_\mu \equiv \exp(im_\mu\delta_\mu) \quad \text{and} \quad D \equiv (9b_{10}b_{23}-4b_{12}b_{21}) \tag{157}$$

the m's are used in specifying the frequencies (138). This definition of C differs slightly from the previous ones in (111) and (79). The value of f^2 is derived from the recursion formula for the A's at j=1 and $n=(\rho-\lambda)\cdot m$. The resulting $C_{\lambda;\rho}$'s will automatically satisfy the resonance condition $n=(\rho-\lambda)\cdot m$ at j=1, in (158). In terms of these C's the linear recursion formula becomes

$$[n-(\rho-\lambda)\cdot m\ \nu_r]f_1^2f_2^2C_{\lambda;\rho}(j,n) = -\sum_{\mu=1}^{2}\{S_\mu[(\lambda_\mu-\rho_\mu)+\lambda_\mu\zeta_\mu^2E_\mu^{-1}F_\mu-\rho_\mu\zeta_\mu^{-2}E_\mu F_\mu^{-1}] + \sum_{(\alpha;\beta)} u_\mu(\alpha;\beta)f_\mu^{-1}[\lambda_\mu\zeta_\mu E_\mu^{-1}-\rho_\mu\zeta_\mu^{-1}F_\mu^{-1}]E^\alpha F^\beta\}C_{\lambda;\rho}(j-1,n) \tag{158}$$

$$\text{with } U_\mu(\alpha;\beta) \equiv u_\mu(\alpha;\beta) \prod_{k=1}^{2} f_k^{\alpha_k+\beta_k}\zeta_k^{\beta_k-\alpha_k} \quad , \tag{159}$$

and where we used an obvious symbolic notation for the E,F powers at the end of **eq.** (148).

Setting **j=0** in (158) we find the $C_{\lambda;\rho}(0,\rho-\lambda)$ to be the only nonvanishing coefficients and, from (155),

$$C_{\lambda;\rho}(0,n) = \delta_{n,(\rho-\lambda)\cdot m} \tag{160}$$

Having obtained these starting values for the C's the higher order coefficients are easily obtained by iteration of our linear recursion relation (158), an operation which can be efficiently programmed for a computing machine. The leading terms in the expansion of $v_\mu(t)$ are

$$v_\mu(t) = f_\mu \exp\{i[m_\mu(\nu_r t+\delta_r)+\delta_\mu]\}+\varepsilon\{..\} + \ldots \tag{161}$$

We now derive, at j=1 and $m\cdot(\rho-\lambda)$, the values of the f^2's listed in (156). In view of the δ form of (160) the only terms in (158) which are proportional to S_1 or S_2, and which do not vanish are the ones with combined forefactor

$$-S\cdot(\lambda-\rho) \tag{162}$$

Among the terms in the sum over the $u(\alpha;\beta)$'s in (158) let us consider the one proportional to λ_1. In view of (160), it has the form

$$-\lambda_1 \sum_{(\alpha;\beta)} u_1(\alpha;\beta) f_1^{-1} \zeta_1 \; \delta_{m\cdot(\beta-\alpha),m_1} \tag{163}$$

The only members of the set $(\alpha;\beta)$ for which the contribution to this sum is non vanishing are those for which

$$m_1(\beta_1-\alpha_1-1) + m_2(\beta_2-\alpha_2) = 0 \ , \tag{164}$$

which can be identified in Table III. Notice that for any m_1 and m_2, eq. (164) is satisfied when

$$(\alpha_1,\beta_1,\alpha_2,\beta_2) = (2,1,0,0) \quad \text{or} \quad (1,0,1,1) \tag{165}$$

There are exceptional cases, i.e. $(m_1,m_2) = (3,1)$ or $(1,3)$ or $(1,1)$, where there are additional $(\alpha;\beta)$ sets which satisfy (164). For example, when $(m_1,m_2) = (3,1)$ the set $(0,0,3,0)$ is appropriate also. We restrict ourselves here to the "normal" case in which (165) constitutes the complete solution of (164).

The exceptional case could be dealt with in a similar manner. The "exceptional cases" can be avoided altogether by scaling the original differential equations (2) and (140) to new variables $\chi(\mu)$ which are identically 1 at t=0, in a manner similar to (135-137) [7]. The special coefficients at $n=m\cdot(\rho-\lambda)$ are then always trivially found from a relation like (83).

In the "normal" case the sum in (163) then becomes, using Table III,

$$\lambda_1(3b_{1,0}f_1^2 + 2b_{1,2}f_2^2) \tag{166}$$

The term proportional to ρ_1 in the $(\alpha;\beta)$ sum of (158) is of a similar form (note the symmetry relation (154) with the $-\lambda_1$ coefficient replaced by ρ_1. Hence the sum of the two terms becomes

$$(\lambda_1-\rho_1)(3b_{1,0}f_1^2 + 2b_{1,2}f_2^2) \qquad (167)$$

The terms proportional to $u_2(\alpha;\beta)$ are treated similarly and one finds the sum of the two, corresponding, nonvanishing terms to be

$$(\lambda_2-\rho_2)(2b_{2,1}f_1^2 + 3b_{2,3}f_2^2) \qquad (168)$$

We are computing special coefficients, i.e. with $n=m\cdot(\rho-\lambda)$, for which the l.h.s. of (158) vanishes. So (162), (167) and (168) must add up to zero and we solve the f's as usual, cf. (108-120), from

$$3b_{1,0}f_1^2 + 2b_{1,2}f_2^2 = S_1 \quad \text{and} \qquad (169)$$

$$2b_{21}f_1^2 + 3b_{2,3}f_2^2 = S_2 \quad , \qquad (170)$$

where S_1 and S_2 are defined in (142), to obtain the values of f_1^2 and f_2^2 "announced" in (156).

We now proceed to calculate the first order coefficients for $n\neq m\cdot(\rho-\lambda)$. At $j=1$ we substitute the zeroth order C's of (160) in the r.h.s. of (158). Let

$$n = (\rho-\lambda + k)\cdot m \quad , \quad \text{with } k \equiv (k_1,k_2)\neq 0 \qquad (171)$$

Then it is found that

$$\nu_r C_{\lambda;\rho}(1,[\rho-\lambda+k]\cdot m) = \sum_{\mu=1}^{2} [\lambda_\mu Q_{\mu,1}(k)+\rho_\mu Q_{\mu,-1}(k)] \quad , \text{with} \qquad (172)$$

$$-s\ k\cdot m\ f_1^2 f_2^2 Q_{\mu s} \equiv S_\mu \zeta_\mu^{2s} \delta_{k\cdot m,2m_\mu s} - P_{\mu s}(k),\ s=-1,+1, \qquad (173)$$

$$P_{\mu s}(k) = f_\mu^{-1} \sum_{(\alpha;\beta)} u_\mu(\alpha;\beta)\ \zeta_\mu^s\ \delta_{(\alpha-\beta+k)\cdot m,m_\mu s} \qquad (174)$$

In the "normal" case the only way of having

$$(\alpha_1-\beta_1+k_1)m_1+(\alpha_2-\beta_2+k_2)m_2 = \delta_{\mu,1}m_1 s+\delta_{\mu,2}m_2 s \qquad (175)$$

is to have same coefficients of m_1, or m_2, on the l.h.s. and r.h.s. of (175), i.e.

$$\alpha_p-\beta_p = s\delta_{p,\mu} - k_p \qquad p=1,2 \qquad (176)$$

Introducing the symbolic notation

$$\zeta^k \equiv \zeta_1^{k_1}\zeta_2^{k_2} \quad \text{and} \quad -k \equiv \{-k_1,-k_2\} \quad , \tag{177}$$

we see that

$$\zeta^k P_{\mu s}(k) = f_\mu^{-1} u_\mu(\alpha;\beta) f_1^{\alpha_1+\beta_1} f_2^{\alpha_2+\beta_2} \delta_{(\alpha-\beta+k)\cdot m, m_\mu s} \tag{178}$$

Using the symmetry relation (154-155) this implies

$$\zeta^{-k} P_{\mu s}(k) = \zeta^k P_{\mu,-s}(-k) \tag{179}$$

in the symbolic notation (177). Combined with (173) this yields

$$\zeta^{-k} Q_{\mu s}(k) = \zeta^k Q_{\mu,-s}(-k) \quad , \tag{180}$$

since in a "normal" case $k \cdot m = 2m_\mu s$ can only hold if $k_p = 2s\delta_{\mu,p}$. Let us now calculate the first order special coefficients, i.e. with $k=\{0,0\}$. At $j=2$ and $k=0$, i.e. $n=m\cdot(\rho-\lambda)$, the l.h.s. of (158) vanishes as usual, cf. (115-116). The r.h.s. contains these first order special C's and we solve for them by transferring them to the l.h.s. of (158), with $n=m\cdot(\rho-\lambda)$. They have the form

$$\begin{aligned} \text{L.H.S.} = \{&(\lambda_1-\rho_1)[S_1-f_1^2u_1(2,1,0,0)E_1F_1-f_2^2u_1(1,0,1,1)E_2F_2] \\ &+ (\lambda_2-\rho_2)[S_2-f_1^2u_2(1,1,1,0)E_1F_1-f_2^2u_2(0,0,2,1)E_2F_2]\} \times \\ &\nu_r C_{\lambda;\rho}(1,m\cdot[\rho-\lambda]) \end{aligned} \tag{181}$$

Again we assume that these C's are linear in the indices, cf. (115-116), and use the symmetry relation (146-154) to obtain

$$C_{\lambda;\rho}(1,m\cdot[\rho-\lambda]) = A(\lambda_1+\rho_1) + B(\lambda_2+\rho_2) \tag{182}$$

Our L.H.S. in (181) then becomes, using Table III and (169-170)

$$\text{L.H.S.}=-2\nu_r\{(\lambda_1-\rho_1)[3f_1^2b_{1,0}A+2f_2^2b_{1,2}B]+(\rho_2-\lambda_2)[2f_1^2b_{2,3}A+3f_2^2b_{2,3}B]\} \tag{183}$$

The r.h.s. of what is left of (158) in this case will be called R.H.S. and has two types of terms, those linear in λ_μ and ρ_μ, the sum of which we will call $(\text{R.H.S.})_l$ and those quadratic in λ_μ and

ρ_μ, the sum of which we call $(\text{R.H.S.})_q$. If (183) is to be valid and yield A and B, which gives us the required Fourier components in (182) then we must be able to prove that

$$(\text{RHS})_q=0 \quad \text{and} \quad (\text{RHS})_l = (\lambda_1-\rho_1)G+(\lambda_2-\rho_2)H, \tag{184}$$

with G and H some proportionality constants. The full expansion of $(\text{RHS})_l$ can be shown to be

$$\begin{aligned}
&S_1\lambda_1\zeta_1^2[Q_{1,1}(-2,0)-Q_{1,-1}(-2,0)]+S_1\rho_1\zeta_1^{-2}[Q_{1,1}(2,0)-Q_{1,-1}(2,0)]\\
&+\lambda_1\Sigma' u_1(\alpha;\beta)f_1^{-1}\zeta_1[(\alpha_1-1)Q_{1,1}(\alpha_1-\beta_1-1,\alpha_2-\beta_2)+\beta_1Q_{1,-1}(\alpha_1-\beta_1^{-1},\alpha_2-\beta_2)\\
&\qquad +\alpha_2Q_{2,1}(\alpha_1-\beta_1-1,\alpha_2-\beta_2)+\beta_2Q_{2,-1}(\alpha_1-\beta_1-1,\alpha_2-\beta_2)]\\
&-\rho_1\Sigma'' u_1(\alpha;\beta)f_1^{-1}\zeta_1^{-1}[\alpha_1Q_{1,1}(\alpha_1-\beta_1+1,\alpha_2-\beta_2)+(\beta_1-1)Q_{1,-1}(\alpha_1-\beta_1+1,\alpha_2-\beta_2)\\
&\qquad +\alpha_2Q_{2,1}(\alpha_1-\beta_1+1,\alpha_2-\beta_2)+\beta_2Q_{2,-1}(\alpha_1-\beta_1+1,\alpha_2-\beta_2)]
\end{aligned} \tag{185}$$

plus corresponding terms with the subscript "1" on λ_1,ρ_1,S_1,u_1, etc. replaced by "2" and the α_2 and β_2 values interchanged with α_1 and β_1. The Σ' implies that the $(\alpha_1,\beta_1,\alpha_2,\beta_2)$ terms (2,1,0,0) and (1,0,1,1) are to be omitted from the summation, Σ'' that (1,2,0,0) and (0,1,1,1) be omitted, Σ''' that (0,0,2,1) and (1,1,1,0) be omitted and Σ'''' that (0,0,1,2) and (1,1,0,1) be omitted. The next step is the use of the symmetries (146), (154) and (178-179) to show that $(\text{RHS})_l$ has the form indicated in (183). Equation (179) implies the symmetry relation

$$\zeta_1^{-2}[Q_{1,1}(2,0)-Q_{1,-1}(2,0)] = \zeta_1^2[Q_{1,1}(-2,0)-Q_{1,-1}(-2,0)] \tag{186}$$

Hence the first two terms in (185) combine in the required manner indicated in (184). The term

$$(-\rho_1 \;\Sigma'' u_1(\alpha;\beta)f_1^{-1}\zeta_1^{-1}(\beta_1-1)Q_{1,-1}(\alpha_1-\beta_1+1,\alpha_2-\beta_2) \tag{187}$$

$$=-\rho_1\Sigma'' U_1(\alpha;\beta)f_1^{\alpha_1+\beta_1-1}f_2^{\alpha_2+\beta_2}\zeta_1^{\beta_1-\alpha_1-1}\zeta_2^{\beta_2-\alpha_2}(\beta_1-1)Q_{1,-1}(\alpha_1-\beta_1+1,\alpha_2-\beta_2)$$

$$=-\rho_1\Sigma'' U_1(\alpha;\beta)f_1^{\alpha_1+\beta_1-1}f_2^{\alpha_2+\beta_2}\zeta_1^{\alpha_1-\beta_1+1}\zeta_2^{\alpha_2-\beta_2}(\alpha_1-1)Q_{1,-1}(\beta_1-\alpha_1-1,\beta_2-\alpha_2)$$

can be shown to have the same form as the first term proportional to λ_1 in the second line of (185) by interchanging β_μ with α_μ for μ=1 and 2 and applying the symmetry property (146). In a similar manner, each term proportional to ρ_1 can be identified with one proportional to λ_1 (and each proportional to ρ_2 with one proportional to λ_2) so that $(RHS)_l$ has the required form

$$(\lambda_1-\rho_1)\{S_1\zeta_1^2[Q_{1,1}(-2,0)-Q_{1,-1}(-2,0)] \qquad (188)$$

$$+\Sigma' u_1(\alpha;\beta)f_1^{-1}\zeta_1[(\alpha_1-1)Q_{1,1}(\alpha_1-\beta_1-1,\alpha_2-\beta_2)+\beta_1 Q_{1,-1}(\alpha_1-\beta_1-1,\alpha_2-\beta_2)$$

$$+\alpha_2 Q_{2,1}(\alpha_1-\beta_1-1,\alpha_2-\beta_2)+\beta_2 Q_{2,-1}(\alpha_1-\beta_1-1,\alpha_2-\beta_2)]\}$$

plus corresponding terms with subscripts 1 on λ_1,ρ_1,S_1,u_1, etc. replaced by "2" and α_2 and β_2 values interchanged with α_1 and β_1.

It now remains to be shown that $(RHS)_q$ vanishes as proposed in (184). The quadratic term proportional to λ_1^2 in $(RHS)_q$ is

$$-\lambda_1^2\{S_1\zeta_1^2 Q_{1,1}(-2,0) - \Sigma' u_1(\alpha;\beta)Q_{1,1}(\alpha_1-\beta_1,\alpha_2-\beta_2)\} \qquad (189)$$

The $Q_{1,1}$ which appears in the sum has two parts, as exhibited in eq. (173). The first part just cancels the first term (that proportional to S_1) in (189). The remaining term has several irrelevant factors in the summand, which when omitted, leaves the following expression which must be shown to vanish

$$\sum_{\{\alpha\}\{\alpha'\}}' U_1(\alpha;\beta)U_1(\alpha';\beta')\{(\alpha_1-\beta_1-1)m_1+(\alpha_2-\beta_2)m_2\}^{-1}\delta_{\alpha_1-\beta_1-1,\beta_1'-\alpha_1'+1}\times$$

$$\delta_{\alpha_2-\beta_2,\beta_2'-\alpha_2'} \qquad (190)$$

One way of showing that this double sum vanishes is to use Table III to sum up all the terms, as we have indeed done. Certain symmetry arguments can be used to obtain the same result. The quadratic term proportional to λ_1,ρ_1 in $(RHS)_q$ can be shown to be

$$\lambda_1\rho_1\{-S_1\zeta_1^2 Q_{1,-1}(-2,0)+\Sigma' u_1(\alpha;\beta)f_1^{-1}\zeta_1 Q_{1,-1}(\alpha_1-\beta_1-1,\alpha_2-\beta_2)$$

$$+S_1\zeta_1^2 Q_{1,1}(2,0)-\Sigma'' u_1(\alpha;\beta)f_1^{-1}\zeta_1^{-1}Q_{1,1}(\alpha_1-\beta_1+1,\alpha_2-\beta_2)\} \qquad (191)$$

A simple application of (146) and (170) can be used to show directly that the sum of the terms in the bracket in (191) vanishes as required. One can easily proceed to show through similar arguments that the other quadratic terms in $(\mathrm{RHS})_q$ vanish.

The required values of the constants A and B which appear in (182) and (183) can then be obtained by comparing the coefficients of $(\lambda_1-\rho_1)$ and $(\lambda_2-\rho_2)$ of (183) and (188) and its equivalent which is proportional to $(\lambda_2-\rho_2)$. One finds

$$-2\nu_r[3b_{1,0}f_1^2\,A+2b_{1,2}f_2^2\,B]$$

$$= \{S_1\zeta_1^2[Q_{1,1}(-2,0)-Q_{1,-1}(-2,0)] \tag{192}$$

$$+\ \Sigma' u_1(\alpha;\beta)f_1^{-1}\zeta_1[(\alpha_1-1)Q_{1,1}(\alpha_1-\beta_1-1,\alpha_2-\beta_2)+\beta_1 Q_{1,-1}(\alpha_1-\beta_1-1,\alpha_2-\beta_2)$$

$$+\alpha_2 Q_{21}(\alpha_1-\beta_1-1,\alpha_2-\beta_2)+\beta_2 Q_{2,-1}(\alpha_1-\beta_1-1,\alpha_2-\beta_2)]\}\ , \quad \text{and}$$

$$-2\nu_r[2b_{2,1}f_1^2\,A+3b_{2,3}f_2^2 B]$$

$$= \{S_2\zeta_2^2[Q_{2,1}(0,-2)-Q_{2,-1}(0,-2)] \tag{193}$$

$$+\ \Sigma'' u_2(\alpha;\beta)f_2^{-1}\zeta_2[(\alpha_1 Q_{1,1}(\alpha_1-\beta_1,\ \alpha_2-\beta_2-1)+\beta_2 Q_{1,2}(\alpha_1-\beta_1,$$

$$\alpha_2-\beta_2-1)$$

$$+(\alpha_2-1)Q_{2,1}(\alpha_1-\beta_1,\alpha_2-\beta_2-1)+\beta_2 Q_{2,2}(\alpha_1-\beta_1,\alpha_2-\beta_2-1)]\}$$

We do not write a final expression here for A and B but merely note that this, in principle, completes the calculation of the first order Fourier coefficients. One can systematically use the recursion formula (158) to find higher order terms. Since the Fourier series are in terms of one basic frequency ν_r, no Small Denominators ever appear. The analysis of some specific examples of coupled oscillator models will be presented in more detail elsewhere[7].

ACKNOWLEDGEMENTS

We would like to thank Mr. Ralph Janda for programming the recursion formulae and obtaining Tables I and II. This research was sponsored in part by the Defense Advanced Research Projects Agency, ARPA, order number 2953 under contract DAAH01-75-C-08031.

REFERENCES

1. C.R. Eminhizer, R.H.G. Helleman and E.W. Montroll, "On a Convergent Nonlinear Perturbation Theory without Small Denominators or Secular Terms", to appear, J. Math. Phys. 17, 121, (1976)
2. T. Carleman, "Application de la Theorie des Equations Integrales Lineaires aux Systemes d'Equations Differentielles Non Lineaires", Acta Mathematica, 59 63-87, (1932); an earlier article on that general subject is in Arkiv for Matematiks, Astronomi o. Fysik, 22B, No. 7, (1931); Uppsala, Sweden.
3. G. Stokes, "On the Theory of Oscillatory Waves," Trans. Cambr. Phil. Soc., VIII, 441, (1847); reprinted in "Mathematics and Physics Papers", p. 197, (1905), Cambridge University Press.
4. M. Lindstedt, Mém. Acad. Roy. Sc. (St. Petersburg), Series 31, 438, (1883).
5. H. Poincaré, "Les Methodes Nouvelles de la Mecanique Celeste" (1892), reprinted by Dover Press (1957).
6. R.H.G.Helleman and E.W. Montroll, "On a Nonlinear Perturbation Theory without Secular Terms, I.", Physica, 74, 22, (1974).
7. C.R. Eminhizer and R.H.G. Helleman, "The Fermi-Pasta-Ulam Chain" in preparation.
8. R. Bellman and J.M. Richardson, Quart. Appl. Math. 20, 333, (1963).
9. C. Ginsburg and R.H.G. Helleman, "Matrix Mechanics of Anharmonic Quantum Oscillators", in preparation.

Dr. Montroll is Einstein Professor of Physics and Chemistry and Director of the Institute for Fundamental Studies at the University of Rochester. He was Lorentz Professor at Leiden, Director of General Science at IBM Technical Center, Vice President for Research at IDA and is a member of the National Academy of Sciences. His recent work includes the theory of turbulence and mathematical modeling of biological and social phenomena. He first met Julius when Julius was a graduate student at Princeton. They collaborated in research on free radical statistics at the National Bureau of Standards.

Dr. Robert Helleman is a Research Associate and Assistant Professor in the Department of Physics and Astronomy at the University of Rochester. His recent work has been in nonlinear dynamics.

GENERALIZED HYDRODYNAMICS

Irwin Oppenheim, Chemistry Department
Mass. Inst. of Technology, Cambridge, Mass. 02139

ABSTRACT

A careful study of the derivation of the non-linear hydrodynamic equations using generalized response theory is presented. There is a correspondence between the results of response theory and the use of local equilibrium distribution functions but it is not direct. The general forms of the transport equations for the densities of conserved variables are presented and discussed.

INTRODUCTION

It seems particularly appropriate to discuss some aspects of generalized hydrodynamics in a volume in memoriam to Julius L. Jackson. Jackson was one of the early workers in the field and made significant contributions to it.[1] He also cast some doubts on the validity of linear response theory.[2] These doubts were later formalized by van Kampen.[3] Indeed, care must be taken in the application of linear response and its generalizations to calculations of the behavior of non-equilibrium systems.

In this paper, we discuss the derivation of non-linear transport equations for a set of macroscopic variables. We shall be particularly interested in how this set of variables relaxes from its initial non-equilibrium values to its equilibrium value. We shall briefly mention the phenomenon of long-time tails[4] which has received much recent attention. These tails are not important in the Navier-Stokes approximation in three dimensions.

RESPONSE THEORY

In this section we discuss the philosophy and results of non-linear response theory. Some caveats in the use of response theory will be mentioned and a comparison of the results of response theory with those obtained from local equilibrium distributions will be presented.

In response theory the system is assumed to be in total equilibrium in the indefinite past, i.e., at $t = -\infty$. Time dependent forces which couple to the microscopic dynamic variables are turned on adiabatically. These forces produce a non-equilibrium distribution function at time $t = 0$. At $t = 0$, the forces are turned off suddenly. The macroscopic properties of the system for $t \geq 0$ are computed using the distribution function at $t = 0$.

The applied forces may be real or fictitious. Their only purpose is to create a distribution function at $t = 0$ which yields the specified initial values of the macroscopic variables. The detailed form of this distribution function is assumed to be unimportant as long as all the slow variables in the system are included in the set of variables whose initial values are specified.

We consider a classical system consisting of N identical particles in a volume V. The dynamical state of the system is determined by specifying the phase point X. The Hamiltonian of the system is H(X) and the distribution function at $t = -\infty$ is

$$f(X,-\infty) = f_0(X) = e^{-\beta H(X)}/\int e^{-\beta H(X)}\, dX \tag{1}$$

where $\beta = (k_B T)^{-1}$, k_B is Boltzmann's constant and T is the absolute temperature.

The set of variables whose average properties we are interested in is denoted by $\underset{\sim}{Q}(X,\underset{\sim}{r})$ which can be written

$$\underset{\sim}{Q}(X,\underset{\sim}{r}) = \sum_{j=1}^{N} \underset{\sim}{Q}_j \delta(\underset{\sim}{r}_j - \underset{\sim}{r}) \tag{2}$$

These quantities are the molecular dynamical variables whose averages are the densities of macroscopic quantities at the space point $\underset{\sim}{r}$. The essential feature of the set $\underset{\sim}{Q}$ is that the time derivatives

$$\frac{d}{dt} \underset{\sim}{Q}(X(t),\underset{\sim}{r}) \equiv \dot{\underset{\sim}{Q}}(X(t),\underset{\sim}{r}) \tag{3}$$

are small so that their averages vary over time intervals which are long compared to molecular times (10^{-13} sec).

The Hamiltonian of the system including the applied forces is given by

$$H_T(X,t) = H(X) - \underset{\sim}{Q}(X,\underset{\sim}{r}) * \underset{\sim}{F}(\underset{\sim}{r},t) \equiv H(X) + H_1(X,t) \tag{4}$$

where

$$\underset{\sim}{Q}(X,\underset{\sim}{r}) * \underset{\sim}{F}(\underset{\sim}{r},t) \equiv \int \underset{\sim}{Q}(X,\underset{\sim}{r}) \cdot \underset{\sim}{F}(\underset{\sim}{r},t)\, d\underset{\sim}{r} \quad , \tag{5}$$

the integral over $\underset{\sim}{r}$ is over the volume V of the system, and the time dependence of the applied forces is

$$\underset{\sim}{F}(\underset{\sim}{r},t) = \begin{cases} \underset{\sim}{F}(\underset{\sim}{r})e^{\varepsilon t}, & \varepsilon \to 0 + t \leq 0 \\ 0 & t > 0 \end{cases} \tag{6}$$

The equation of motion for the distribution function is

$$\frac{\partial f(X,t)}{\partial t} = -iL_T(t) f(X,t) \tag{7}$$

where the Liouville operator $iL_T(t)$ is given by

$$iL_T(t) = iL + iL_1(t); \tag{8}$$

iL is the Liouville operator corresponding to H and iL_1 is the Liouville operator corresponding to H_1. The solution of Eq. (7) is

$$f(X,t) = \exp \{- \int_{-\infty}^{t} iL_T(\tau)d\tau\}_+ \; f(X,-\infty) \tag{9}$$

where the subscript + implies a time ordering operation.

The expansion of Eq. (9) to second order in $\mathbf{F}$ is

$$f(X,t) = f_0(X) + \int_{-\infty}^{t} d\sigma e^{-iL(t-\sigma)}[-iL_1(\sigma)]f_0(X)$$

$$+ \int_{-\infty}^{t} d\sigma \int_{-\infty}^{\sigma} d\sigma^1 e^{-iL(t-\sigma)}[-iL_1(\sigma)]e^{-iL(\sigma-\sigma^1)}[-iL_1(\sigma^1)]f_0$$

$$\equiv f_0(X) + \int_{-\infty}^{t} d\sigma e^{-iL(t-\sigma)}[f_0(X), \mathbf{Q}(X,\mathbf{r})] * \mathbf{F}(\mathbf{r},\sigma)$$

$$+ \int_{-\infty}^{t} d\sigma \int_{-\infty}^{\sigma} d\sigma^1 e^{-iL(t-\sigma)}[e^{-iL(\sigma-\sigma^1)}[f_0(X), \mathbf{Q}(X,\mathbf{r})]$$

$$* \mathbf{F}(\mathbf{r},\sigma^1), \mathbf{Q}(X,\mathbf{r}^1)] * \mathbf{F}(\mathbf{r}^1,\sigma) \tag{10}$$

where the symbol [,] denotes the Poisson bracket. We make use of the fact that

$$[f_0, \mathbf{Q}] = \beta\dot{\mathbf{Q}}f_0 \tag{11}$$

to rewrite Eq. (10) as:

$$f(X,t) = f_0(X) + \beta\int_{-\infty}^{t} d\sigma\dot{\mathbf{Q}}(\mathbf{r},\sigma-t) * \mathbf{F}(\mathbf{r},\sigma)f_0(X)$$

$$+ \beta\int_{-\infty}^{t} d\sigma\int_{-\infty}^{\sigma} d\sigma^1\{\beta\dot{\mathbf{Q}}(\mathbf{r},\sigma^1-t) * \mathbf{F}(\mathbf{r},\sigma^1)\dot{\mathbf{Q}}(\mathbf{r}^1,\sigma-t) * \mathbf{F}(\mathbf{r}^1,\sigma)f_0(X)$$

$$+ e^{-iL(t-\sigma)}[\dot{\mathbf{Q}}(\mathbf{r},\sigma^1-\sigma), \mathbf{Q}(\mathbf{r}^1)]f_0(X) \mathbin{\substack{*\\ *}} \mathbf{F}(\mathbf{r},\sigma^1)\mathbf{F}(\mathbf{r}^1,\sigma)\} \tag{12}$$

where we have used the abbreviated notation $\mathbf{Q}(X,\mathbf{r}) \equiv \mathbf{Q}(\mathbf{r})$ and $\mathbf{Q}(X(t),\mathbf{r}) \equiv \mathbf{Q}(\mathbf{r},t)$. The expansion in Eq. (12) preserves the normalization of $f(X,t)$.

Equation (12) is clearly an unsuitable expansion for $f(X,t)$. Since $\mathbf{Q}$ contains N terms it is an expansion in NF rather than an expansion in powers of F. This is consistent with the objections raised by Jackson[2] and van Kampen[3] and arises due to the fact that f is an N body function. This does not, however, imply that expansions similar to Eq. (12) will not be appropriate for reduced distribution functions or for macroscopic variables.

We use the fact that the average of a dynamical variable for $t \geq 0$ can be computed using the distribution function at $t = 0$. We define $\mathbf{q}(\mathbf{r},t)$ by

$$\mathbf{q}(\mathbf{r},t) \equiv \int f(X,0)\mathbf{Q}(\mathbf{r},t)dX - \langle\mathbf{Q}(\mathbf{r},t)\rangle \tag{13}$$

where the symbol < > denotes an average over the equilibrium distribution function $f_0(X)$, i.e.,

$$<Q(\underset{\sim}{r},t)> \equiv \int Q(\underset{\sim}{r},t) f_0(X)\,dX = <Q(\underset{\sim}{r})> \; . \tag{14}$$

Thus,

$$q(\underset{\sim}{r},t) \equiv \int \hat{Q}(\underset{\sim}{r},t) f(X,0)\,dX \; , \tag{15}$$

where

$$\hat{Q}(\underset{\sim}{r},t) \equiv Q(\underset{\sim}{r},t) - <Q(\underset{\sim}{r})> \; , \tag{16}$$

and $q(\underset{\sim}{r},t)$ is the average deviation from equilibrium of the density at point $\underset{\sim}{r}$ of some macroscopic property.

Substitution of Eq. (9) into Eq. (13) and carrying through an expansion to second order in F yields:

$$\begin{aligned} q(\underset{\sim}{r},t) &= \beta \int_{-\infty}^{0} d\sigma <\hat{Q}(\underset{\sim}{r},t)\dot{\hat{Q}}(\underset{\sim}{r}^1,\sigma)> * \tilde{F}(\underset{\sim}{r}^1,\sigma) \\ &+ \beta \int_{-\infty}^{0} d\sigma \int_{-\infty}^{\sigma} d\sigma^1 \{\beta <\hat{Q}(\underset{\sim}{r},t)\dot{\hat{Q}}(\underset{\sim}{r}^1,\sigma^1)\hat{Q}(\underset{\sim}{r}^{11},\sigma)> \overset{*}{*} F(\underset{\sim}{r}^{11},\sigma)F(\underset{\sim}{r},\sigma^1) \\ &+ <\hat{Q}(\underset{\sim}{r},t-\sigma)[\dot{\hat{Q}}(\underset{\sim}{r}^1,\sigma^1-\sigma),\hat{Q}(\underset{\sim}{r}^{11})]> \overset{*}{*} F(\underset{\sim}{r}^{11},\sigma)F(\underset{\sim}{r}^1,\sigma^1) . \end{aligned} \tag{17}$$

Most of the time integrations in Eq. (17) can be carried out explicitly because of the simple form of the time dependence of the F's, Eq. (6). The result is

$$\begin{aligned} q(\underset{\sim}{r},t) &= \beta\{<\hat{Q}(\underset{\sim}{r},t)\hat{Q}(\underset{\sim}{r}^1)> - \lim_{\sigma\to-\infty}<\hat{Q}(\underset{\sim}{r},t)\hat{Q}(\underset{\sim}{r}^1,\sigma)>\} * F(\underset{\sim}{r}^1) \\ &+ \frac{\beta^2}{2}\{<\hat{Q}(\underset{\sim}{r},t)\hat{Q}(\underset{\sim}{r}^1)\hat{Q}(\underset{\sim}{r}^{11})> - 2\lim_{\sigma^1\to-\infty}<\hat{Q}(\underset{\sim}{r},t)\hat{Q}(\underset{\sim}{r}^1,\sigma^1)\hat{Q}(\underset{\sim}{r}^{11})> \\ &- \lim_{\sigma\to-\infty}<\hat{Q}(\underset{\sim}{r},t)\hat{Q}(\underset{\sim}{r}^1,\sigma)\hat{Q}(\underset{\sim}{r}^{11},\sigma)> \\ &+ 2\lim_{\sigma\to-\infty}\lim_{\sigma^1\to-\infty}<\hat{Q}(\underset{\sim}{r},t)\hat{Q}(\underset{\sim}{r}^1,\sigma^1)\hat{Q}(\underset{\sim}{r}^{11},\sigma)>\} \overset{*}{*} F(\underset{\sim}{r}^{11})F(\underset{\sim}{r}^1) \\ &- \beta\int_{\infty}^{0} d\sigma\, e^{2\varepsilon\sigma}<\hat{Q}(\underset{\sim}{r},t-\sigma)[\hat{Q}(\underset{\sim}{r}^1,\sigma^1-\sigma),\hat{Q}(\underset{\sim}{r}^{11})]>\} \overset{*}{*} F(\underset{\sim}{r}^{11})F(\underset{\sim}{r}^1) . \end{aligned} \tag{18}$$

The time limits in Eq. (18) are readily evaluated by using the following considerations. The correlation function $<AB(t)>$ factorizes as $t \to \underline{+}\infty$ unless either A or B are constants of the motion, i.e., iLA or iLB equals zero. Any dynamical variable can be written as

$$B(\underset{\sim}{r},t) \equiv PB(\underset{\sim}{r},t) + (1-P)B(\underset{\sim}{r},t) \tag{19}$$

where P is a projection operator onto the set of conserved variables $\underset{\sim}{C}$, i.e.,

$$PB(\underset{\sim}{r},t) \equiv \langle B(\underset{\sim}{r})\underset{\sim}{C}\rangle \cdot \langle \underset{\sim}{C}\underset{\sim}{C}\rangle^{-1} \cdot \underset{\sim}{C} = PB(\underset{\sim}{r}). \tag{20}$$

Equation (18) can then be rewritten

$$\begin{aligned} q(\underset{\sim}{r},t) &= \beta\{\langle\hat{Q}(\underset{\sim}{r},t)\hat{Q}(\underset{\sim}{r}^1)\rangle - \langle\hat{Q}(\underset{\sim}{r})P\hat{Q}(\underset{\sim}{r}^1)\rangle\} * F(\underset{\sim}{r}^1) \\ &+ \frac{\beta^2}{2}\{\langle\hat{Q}(\underset{\sim}{r},t)\hat{Q}(\underset{\sim}{r}^1)\hat{Q}(\underset{\sim}{r}^{11})\rangle - 2\langle\hat{Q}(\underset{\sim}{r},t)P\hat{Q}(\underset{\sim}{r}^1)\hat{Q}(\underset{\sim}{r}^{11})\rangle \\ &- \langle(P\hat{Q}(\underset{\sim}{r}))\hat{Q}(\underset{\sim}{r}^1)\hat{Q}(\underset{\sim}{r}^{11})\rangle + 2\langle\hat{Q}(\underset{\sim}{r})(P\hat{Q}(\underset{\sim}{r}^1))P\hat{Q}(\underset{\sim}{r}^{11})\rangle\} \overset{*}{*} F(\underset{\sim}{r}^{11})F(\underset{\sim}{r}^1) \\ &- \beta\int_{-\infty}^{0} d\sigma e^{2\varepsilon\sigma}\langle(1-P)\hat{Q}(\underset{\sim}{r},t-\sigma)[P\hat{Q}(\underset{\sim}{r}^1),(1-P)\hat{Q}(\underset{\sim}{r}^{11})]\rangle \overset{*}{*} F(\underset{\sim}{r}^{11})F(\underset{\sim}{r}^1). \end{aligned} \tag{21}$$

The time integral in the last term in Eq. (21) is finite since

$$\begin{aligned} &\lim_{\sigma\to-\infty} \langle Q(\underset{\sim}{r},t)[P\hat{Q}(\underset{\sim}{r}^1),\hat{Q}(\underset{\sim}{r}^{11},\sigma)]\rangle \overset{*}{*} F(\underset{\sim}{r}^{11})F(\underset{\sim}{r}^1) \\ &= \langle Q(\underset{\sim}{r},t)[P\hat{Q}(\underset{\sim}{r}^1),P\hat{Q}(\underset{\sim}{r}^{11})]\rangle \overset{*}{*} F(\underset{\sim}{r}^{11})F(\underset{\sim}{r}^1) \\ &= 0 . \end{aligned} \tag{22}$$

The coefficient of $\underset{\sim}{F}$ in Eq. (21) is of order N^0 as is the coefficient of $\underset{\sim}{F}^2$. Thus, Eq. (21) does provide a suitable expansion of $q(\underset{\sim}{r},t)$ in powers of $\underset{\sim}{F}$.

Finally, Eq. (21) can be rewritten:

$$\begin{aligned} q(\underset{\sim}{r},t) &= \beta\langle(1-P)\hat{Q}(\underset{\sim}{r},t)(1-P)\hat{Q}(\underset{\sim}{r}^1)\rangle * F(\underset{\sim}{r}^1) \\ &+ (\beta^2/2)\{\langle[(1-P)\hat{Q}(\underset{\sim}{r},t)][(1-P)\hat{Q}(\underset{\sim}{r}^1)](1-P)\hat{Q}(\underset{\sim}{r}^{11})\rangle \overset{*}{*} F(\underset{\sim}{r}^{11})F(\underset{\sim}{r}^1) \\ &- \beta\int_{-\infty}^{0} d\sigma e^{2\varepsilon\sigma}\langle[(1-P)Q(\underset{\sim}{r},t-\sigma)][P\hat{Q}(\underset{\sim}{r}^1),(1-P)\hat{Q}(\underset{\sim}{r}^{11})]\rangle \overset{*}{*} F(\underset{\sim}{r}^{11})F(\underset{\sim}{r}^1). \end{aligned} \tag{23}$$

If $Q(\underset{\sim}{r},t)$ is the density of a conserved variable, then $\int d\underset{\sim}{r}Q(\underset{\sim}{r},t)$ is a conserved variable and it follows from Eq. (23) that

$$\int_V q(\underset{\sim}{r},t)d\underset{\sim}{r} = 0 \tag{24}$$

for conserved variables. Thus, the applied forces do not affect the total number of particles, energy or momentum of the system and the final equilibrium state to which the system relaxes as $t \to \infty$ is identical to that as $t \to -\infty$ as far as measured properties are concerned. The N particle distribution function need not approach f_0 as $t \to +\infty$.

We now specialize to the case in which the equilibrium system is uniform. The set $\mathbf{Q}(\mathbf{r})$ consists of densities of conserved variables. Equation (23) then simplifies considerably if we choose the forces which couple to these variables in such a way that $\int_V \mathbf{F}(\mathbf{r})d\mathbf{r} = 0$. The result is that

$$\mathbf{q}(\mathbf{r},t) = \beta\langle\hat{\mathbf{Q}}(\mathbf{r},t)\hat{\mathbf{Q}}(\mathbf{r}^1)\rangle * \mathbf{F}(\mathbf{r}^1)$$

$$+ (\beta^2/2)\langle[(1-P)\hat{\mathbf{Q}}(\mathbf{r},t)]\hat{\mathbf{Q}}(\mathbf{r}^1)\hat{\mathbf{Q}}(\mathbf{r}^{11})\rangle \overset{*}{*} \mathbf{F}(\mathbf{r}^{11})\mathbf{F}(\mathbf{r}^1). \tag{25}$$

Note that Eq. (25) is <u>not</u> of the form that would arise from a local equilibrium ensemble at t = 0 of the form:

$$f(X,0) = e^{-\beta H_T(X,0)} / \int e^{-\beta H_T(X,0)} dX \tag{26}$$

However, as we shall see later, this difference is not important in the derivation of the hydrodynamic equations.

The transformation from Eq. (23) to Eq. (25) is an important one for the derivations of the hydrodynamic equations carried out in the next section. There we will express $\mathbf{F}$ in terms of $\mathbf{q}$ and will use the inverse of the coefficient of the linear term of the expression for $\mathbf{q}$ in terms of $\mathbf{F}$ and $\mathbf{FF}$. The coefficient in Eq. (23) does not have an inverse nor does the coefficient in Eq. (25) as it stands. We will therefore perform some additional transformations on Eq. (25) before ending this section.

The correlation function $\langle\hat{\mathbf{Q}}(\mathbf{r},t)\hat{\mathbf{Q}}(\mathbf{r}^1)\rangle$ is evaluated using the canonical distribution function. These correlation functions have two unpleasant features. First, they do not factorize[5] for $\mathbf{r}^1$ far from $\mathbf{r}$, i.e., for $|\mathbf{r}-\mathbf{r}^1|\gg\lambda$, where λ is a mean free path, and in fact

$$\langle\hat{\mathbf{Q}}(\mathbf{r},t)\hat{\mathbf{Q}}(\mathbf{r}^1)\rangle \to \frac{\mathbf{C}(t)}{N} . \tag{27}$$

Second, if $\hat{Q}(\mathbf{r},t)$ is the dynamical variable corresponding to the number density, i.e.,

$$\hat{Q}(\mathbf{r},t) = \sum_{j=1}^{N} \delta(\mathbf{r}_j(t)-\mathbf{r}) - N/V \equiv \hat{N}(\mathbf{r},t), \tag{28}$$

$$\text{then } \int_V \hat{Q}(\mathbf{r},t)d\mathbf{r} = 0 \tag{29}$$

and the matrix $\langle\hat{\mathbf{Q}}(\mathbf{r},t)\hat{\mathbf{Q}}(\mathbf{r}^1)\rangle$ does not have an inverse. We can rewrite the linear term in Eq. (25) as

$$\langle\hat{\mathbf{Q}}(\mathbf{r},t)\hat{\mathbf{Q}}(\mathbf{r}^1)\rangle * \mathbf{F}(\mathbf{r}^1) = \int d\mathbf{r}^1\{\langle\hat{\mathbf{Q}}(\mathbf{r},t)\hat{\mathbf{Q}}(\mathbf{r}^1)\rangle - \frac{\mathbf{C}(t)}{N}\} * \mathbf{F}(\mathbf{r}) \tag{30}$$

which follows since $\int_V F(r)dr = 0$. The term on the RHS of Eq. (30) can be rewritten as

$$\int_V dr^1 \langle Q(r,t)Q(r^1)\rangle_{GC} \cdot F(r^1) + 0(\frac{1}{N}) \tag{31}$$

where the subscript GC implies that the correlation function is evaluated using the grand canonical ensemble. Eq. (31) follows from Eq. (30) since the grand canonical average does factorize exactly for large $|r - r^1|$ and the grand canonical average differs from the canonical average by terms of $0(1/N)$. Furthermore, in the grand canonical ensemble, Eq. (28) becomes

$$\int_V \hat{N}(r,t)dr = N - \langle N\rangle_{GC} , \tag{32}$$

and the matrix $\langle\hat{Q}(r,t)\hat{Q}(r^1)\rangle_{GC}$ does have an inverse.

Finally, we rewrite Eq. (25) correct to terms of $0(1/N)$ as

$$q(r,t) = \beta\langle\hat{Q}(r,t)\hat{Q}(r^1)\rangle_{GC} * F(r^1) + (\frac{\beta^2}{2})\langle[(1-P)Q(r,t)]\hat{Q}(r^1)\hat{Q}(r^{11})\rangle \overset{*}{*} F(r^{11})F(r^1) \tag{33}$$

TRANSPORT EQUATIONS

In this section, we derive the hydrodynamic equations using the results of the previous section. The techniques used are similar to those of Weare and Oppenheim.[6]

The basic equations are:

$$q(r,t) = \beta K(r,r^1;t) * F(r^1) + (\frac{\beta^2}{2})L(r,r^1,r^{11};t) \overset{*}{*} F(r^{11})F(r^1) \tag{34}$$

and

$$\dot{q}(r,t) = \beta\dot{K}(r,r^1;t) * F(r^1) + (\frac{\beta^2}{2})\dot{L}(r,r^1,r^{11};t) \overset{*}{*} F(r^{11})F(r^1). \tag{35}$$

Here, it follows from Eq. (33) that

$$K(r,r^1;t) \equiv \langle\hat{Q}(r,t)\hat{Q}(r^1)\rangle_{GC} \tag{36}$$

and

$$L(r,r^1,r^{11};t) \equiv \langle[(1-P)\hat{Q}(r,t)]\hat{Q}(r^1)\hat{Q}(r^{11})\rangle . \tag{37}$$

These equations are also correct with the ^'s removed from any or all of the Q's.

We now invert Eq. (34) to find F in terms of q(r,t) and substitute the result into Eq. (35). The results are:

$$\beta F(r) = K^{-1}(r,r^1;t) * q(r^1,t) - \frac{1}{2}K^{-1}(r,r^1;t) * L(r^1,r^{11},r^{(3)};t) \overset{*}{*} K^{-1}(r^{(3)},r^{(4)};t) * q(r^{(4)},t)K^{-1}(r^{11},r^{(5)};t) * q(r^{(5)},t), \tag{38}$$

and

$$\dot{q}(r,t) = \dot{K}(r,r^1;t) * K^{-1}(r^1,r^{11};t) * q(r^{11},t) +$$
$$\frac{1}{2}\{\dot{L}(r,r^1,r^{11};t) \overset{*}{*} K^{-1}(r^{11},r^{(3)};t) * q(r^{(3)};t)K^{-1}(r^1,r^{(4)};t) * q(r^{(4)},t)$$
$$- \dot{K}(r,r^1;t) * K^{-1}(r^1,r^{11};t) * L(r^{11},r^{(3)},r^{(4)};t) \overset{*}{*} K^{-1}(r^{(4)},r^{(5)};t)$$
$$* q(r^{(5)},t)K^{-1}(r^{(3)},r^{(6)};t) * q(r^{(6)},t)\} \tag{39}$$

Equation (39) is our fundamental equation. We rewrite this equation as:

$$\dot{q}(r,t) = M(r,r^1;t) * q(r^1,t) + N(r,r^1,r^{11};t) \overset{*}{*} q(r^{11},t)q(r^1,t), \tag{40}$$

where the definitions of M and N follow from Eq. (39).

Since all of the variables Q are slowly varying we recast Eq. (40) by making as many time derivatives as possible explicit. Thus,

$$\dot{q}(r,t) = M(r,r^1;0) * q(r^1,t) + N(r,r^1,r^{11};0) \overset{*}{*} q(r^{11},t)q(r^1,t)$$
$$+ \int_0^t \frac{d}{d\tau}\{M(r,r^1;\tau) * q(r^1,t) + N(r,r^1,r^{11};\tau) \overset{*}{*} q(r^{11},t)q(r^1,t)\}d\tau. \tag{41}$$

The first two terms on the RHS of Eq. (41) each contain one time derivative; they correspond to the Euler terms in hydrodynamics and contain one smallness factor. The terms in the integral contain two time derivatives; they correspond to the dissipative (Navier-Stokes) terms and contain at least two smallness factors.

The linear term in the integral in Eq. (41) correct to second order in the smallness parameters characterizing $\dot{Q}$ becomes:

$$\{\ddot{K}(r,r^1;\tau) * K^{-1}(r^1,r^{11};0) - \dot{K}(r,r^1;0) * K^{-1}(r^1,r^{(3)};0)$$
$$* \dot{K}(r^{(3)},r^{(4)};0) * K^{-1}(r^{(4)},r^{11};0)\} * q(r^{11},t)$$
$$= -\langle I(r,\tau)I(r^1)\rangle * K^{-1}(r^1,r^{11};0) * q(r^{11},t), \tag{42}$$

where we have used the fact that

$$\dot{K}^{-1} = -K^{-1} * \dot{K} * K^{-1}, \tag{43}$$

and I is defined by

$$I(r,\tau) = \dot{Q}(r,\tau) - M(r,r^1;0) * \hat{Q}(r^1) . \tag{44}$$

The quantity I is a generalized dissipative current which has the property that

$$<\underset{\sim}{I}(\underset{\sim}{r},\tau)\hat{Q}(\underset{\sim}{r}^{11})> = <\dot{Q}(\underset{\sim}{r},\tau)\hat{Q}(\underset{\sim}{r}^{11})> - <\dot{Q}(\underset{\sim}{r})\hat{Q}(\underset{\sim}{r}^{(3)})>$$

$$* <\hat{Q}(\underset{\sim}{r}^{(3)})\hat{Q}(\underset{\sim}{r}^{(4)})>^{-1} * <\hat{Q}(\underset{\sim}{r}^{(4)})\hat{Q}(\underset{\sim}{r}^{11})> = 0 \quad , \tag{45}$$

to first order in the smallness parameter.

The non-linear term in the integral in Eq. (41) correct to second order becomes:

$$+ \frac{1}{2} <\underset{\sim}{I}(\underset{\sim}{r},\tau)\underset{\sim}{I}(\underset{\sim}{r}^{1})> * \underset{\approx}{K}^{-1}(\underset{\sim}{r}^{1},\underset{\sim}{r}^{11};0) * <\hat{Q}(\underset{\sim}{r}^{11})\hat{Q}(\underset{\sim}{r}^{(3)})\hat{Q}(\underset{\sim}{r}^{(4)})>$$

$$\overset{*}{*} \underset{\approx}{K}^{-1}(\underset{\sim}{r}^{(4)},\underset{\sim}{r}^{(5)};0) * \underset{\sim}{q}(\underset{\sim}{r}^{(5)},t)\underset{\approx}{K}^{-1}(\underset{\sim}{r}^{(3)},\underset{\sim}{r}^{(6)};0) * \underset{\sim}{q}(\underset{\sim}{r}^{(6)},t)$$

$$- \frac{1}{2} \{<\underset{\sim}{I}(\underset{\sim}{r},\tau)\underset{\sim}{I}(\underset{\sim}{r}^{1})\hat{Q}(\underset{\sim}{r}^{11})> + <\underset{\sim}{I}(\underset{\sim}{r},\tau)\hat{Q}(\underset{\sim}{r}^{1})\underset{\sim}{I}(\underset{\sim}{r}^{11})>\}$$

$$\overset{*}{*} \underset{\approx}{K}^{-1}(\underset{\sim}{r}^{11},\underset{\sim}{r}^{(3)};0) * \underset{\sim}{q}(\underset{\sim}{r}^{(3)},t)\underset{\approx}{K}^{-1}(\underset{\sim}{r}^{1},\underset{\sim}{r}^{(4)};0)\underset{\sim}{q}(\underset{\sim}{r}^{(4)},t) . \tag{46}$$

The terms in L involving $P\hat{\underset{\sim}{Q}}$ disappear because of the term involving $\dot{\underset{\approx}{K}}(0) * \underset{\approx}{K}^{-1}(0)$ * in front of these terms. This term becomes

$$\int \dot{\underset{\approx}{K}}(\underset{\sim}{r},\underset{\sim}{r}^{1};0)\underset{\approx}{K}^{-1}(\underset{\sim}{r}^{1},\underset{\sim}{r}^{11};0)d\underset{\sim}{r}^{1}d\underset{\sim}{r}^{11}$$

$$= \int <\dot{\hat{Q}}(\underset{\sim}{r})\hat{Q}(\underset{\sim}{r}^{1})>d\underset{\sim}{r}^{1} * \int \underset{\approx}{K}^{-1}(\underset{\sim}{r}^{1},\underset{\sim}{r}^{11};0)d\underset{\sim}{r}^{11} \quad . \tag{47}$$

The first integral on the RHS is clearly zero. The correlation function $<\hat{Q}(\underset{\sim}{r})\hat{Q}(\underset{\sim}{r}^{1})\hat{Q}(\underset{\sim}{r}^{11})>$ again does not factorize when any of the distances $|\underset{\sim}{r}-\underset{\sim}{r}^{1}|$, $|\underset{\sim}{r}-\underset{\sim}{r}^{11}|$ or $|\underset{\sim}{r}^{1}-\underset{\sim}{r}^{11}|$ become large because it is a canonical average. However, if $\underset{\sim}{r}^{1}$ and $\underset{\sim}{r}^{11}$ are far from $\underset{\sim}{r}$, the function is independent of $\underset{\sim}{r}$ and the terms involving $\underset{\approx}{L}$ in Eq. (39) are zero for the same reason as in the last paragraph. If $\underset{\sim}{r}$ and $\underset{\sim}{r}^{11}$ are far from $\underset{\sim}{r}^{1}$, the function becomes independent of $\underset{\sim}{r}^{1}$ and the term yields zero because of the integrations of $\underset{\approx}{K}^{-1} * \underset{\sim}{q}$ and the fact that $\int \underset{\sim}{q}\, d\underset{\sim}{r} = 0$. A similar argument leads to the result that the term yields zero if $\underset{\sim}{r}$ and $\underset{\sim}{r}^{1}$ are far from $\underset{\sim}{r}^{11}$. Thus, again we can replace the correlation function $<\hat{Q}(\underset{\sim}{r})\hat{Q}(\underset{\sim}{r}^{1})\hat{Q}(\underset{\sim}{r}^{11})>$ by a grand canonical average.

The time integral in Eq. (41) can be extended to infinity because of the properties of the dissipative currents $\underset{\sim}{I}$. Since the $\underset{\sim}{I}$'s are orthogonal to the $\hat{Q}(\underset{\sim}{r})$, they can be assumed, with some reservations, to vary on a molecular time scale. Thus the correlation functions $<\underset{\sim}{I}(\tau)\underset{\sim}{I}>$ and $<\underset{\sim}{I}(\tau)\underset{\sim}{I}\hat{\underset{\sim}{Q}}>$ decay to zero before the slow variables change appreciably. The reservations expressed above are due to the fact that $\underset{\sim}{I}$ is not orthogonal to the non-linear quantities

$\mathbf{QQ}$, $\mathbf{QQQ}$, etc. However, as has been demonstrated elsewhere,[4] this is not important to Navier-Stokes order (i.e., to second order in the smallness parameters characterizing $\mathbf{Q}$ in three dimensions.

We have seen that the relation of $\mathbf{q}(\mathbf{r},t)$ to the forces, $\mathbf{F}(\mathbf{r})$, given by response theory is not of the form that can be obtained from a local equilibrium distribution function. This is because of the coefficient $\mathbf{L}$ of the quadratic term. However, we have also seen that the part of $\mathbf{L}$ which contributes to the transport equations, Eq. (41), is of the form

$$\mathbf{L}'(\mathbf{r},\mathbf{r}',\mathbf{r}'';t) = \langle \hat{\mathbf{Q}}(\mathbf{r},t)\hat{\mathbf{Q}}(\mathbf{r}')\hat{\mathbf{Q}}(\mathbf{r}'')\rangle_{GC} \,. \tag{48}$$

This will be important for our future developments.

Eq. (41) is simplified considerably if we define time dependent force-like quantities, $\boldsymbol{\Phi}(\mathbf{r},t)$, by the equation:

$$\mathbf{q}(\mathbf{r},t) \equiv \beta\mathbf{K}(\mathbf{r},\mathbf{r}';0) * \boldsymbol{\Phi}(\mathbf{r}',t) + \left(\frac{\beta^2}{2}\right)\mathbf{L}'(\mathbf{r},\mathbf{r}',\mathbf{r}'';0) * \boldsymbol{\Phi}(\mathbf{r}'',t)\boldsymbol{\Phi}(\mathbf{r}',t). \tag{49}$$

Then Eq. (41) becomes:

$$\dot{\mathbf{q}}(\mathbf{r},t) = \beta\langle\dot{\hat{\mathbf{Q}}}(\mathbf{r})\hat{\mathbf{Q}}(\mathbf{r}')\rangle * \boldsymbol{\Phi}(\mathbf{r}',t) + \frac{\beta^2}{2}\langle\dot{\hat{\mathbf{Q}}}(\mathbf{r})\hat{\mathbf{Q}}(\mathbf{r}')\hat{\mathbf{Q}}(\mathbf{r}'')\rangle * \boldsymbol{\Phi}(\mathbf{r}'',t)\boldsymbol{\Phi}(\mathbf{r}',t)$$
$$- \int_0^\infty \{\beta\mathbf{I}(\mathbf{r},\tau)\mathbf{I}(\mathbf{r}')\rangle * \boldsymbol{\Phi}(\mathbf{r}') + \left(\frac{\beta^2}{2}\right)[\langle\mathbf{I}(\mathbf{r},\tau)\mathbf{I}(\mathbf{r}')\hat{\mathbf{Q}}(\mathbf{r}'')\rangle$$
$$+ \langle\mathbf{I}(\mathbf{r},\tau)\hat{\mathbf{Q}}(\mathbf{r}')\mathbf{I}(\mathbf{r}'')\rangle] * \boldsymbol{\Phi}(\mathbf{r}'',\tau)\boldsymbol{\Phi}(\mathbf{r}',\tau)\}d\tau \quad . \tag{50}$$

These equations can be written in a suggestive form by recognizing the fact that Eq. (49) is the first two terms in the expansion of the average of $\hat{\mathbf{Q}}(\mathbf{r})$ over the time dependent grand canonical local equilibrium function

$$f_L(X,t) = \frac{f_{GC}(X)\,e^{\beta\mathbf{Q}(\mathbf{r}') * \boldsymbol{\Phi}(\mathbf{r}',t)}}{\sum_{N=0}^{\infty}\int dX\, f_{GC}(X)\,e^{\beta\mathbf{Q}(\mathbf{r}') * \boldsymbol{\Phi}(\mathbf{r}',t)}} \,. \tag{51}$$

Averages performed with this distribution function will be denoted by $\langle\ \rangle_L(t)$. Thus Eq. (49) becomes

$$\mathbf{q}(\mathbf{r},t) = \langle\hat{\mathbf{Q}}(\mathbf{r})\rangle_L(t) \tag{52}$$

and Eq. (50) becomes

$$\dot{\mathbf{q}}(\mathbf{r},t) = \langle\dot{\hat{\mathbf{Q}}}(\mathbf{r})\rangle_L(t) - \beta\int_0^\infty \langle\mathbf{I}(\mathbf{r},\tau)\mathbf{I}(\mathbf{r}')\rangle_L(t) * \boldsymbol{\Phi}(\mathbf{r}',t)d\tau \,, \tag{53}$$

where we have used the fact that $\langle\dot{\mathbf{Q}}(\mathbf{r})\rangle = 0$.

The significance of the $\boldsymbol{\Phi}$'s can now readily be seen. They are the conjugate forces to the macroscopic variables $q(\tilde{r},t)$.[6] The set of the hydrodynamic variables consists of the number density, $n(\tilde{r},t)$, the momentum density, $\tilde{p}(\tilde{r},t)$, and the energy density, $e(\tilde{r},t)$. The corresponding conjugate forces are:

$$\Phi_N(\tilde{r},t) = [\beta(\tilde{r},t)\phi(\tilde{r},t) - \beta\mu]/\beta \tag{54}$$

where $\beta(\tilde{r},t)$ is a space and time dependent inverse temperature,

$$\phi(\tilde{r},t) = \tilde{\mu}(\tilde{r},t) - \frac{1}{2}mv^2(\tilde{r},t) \tag{55}$$

is the time and space dependent chemical potential $\mu(r,t)$ minus the local kinetic energy, and $\tilde{v}(\tilde{r},t)$ is the velocity of the system at $\tilde{r}$ and t, defined by

$$\tilde{p}(\tilde{r},t) = mn(\tilde{r},t)v(\tilde{r},t), \tag{56}$$

$$\tilde{\Phi}_p(\tilde{r},t) = [\beta(\tilde{r},t)/\beta]\tilde{v}(\tilde{r},t), \tag{57}$$

and

$$\Phi_E(\tilde{r},t) = -[\beta(\tilde{r},t)-\beta]/\beta \ . \tag{58}$$

Next, we define the set of currents, $\tilde{J}(\tilde{r})$, by

$$\dot{\tilde{Q}}(\tilde{r}) = -\nabla_r \cdot \tilde{J}(\tilde{r}) \tag{59}$$

and the set of dissipative currents, $\tilde{H}(\tilde{r})$, by

$$\tilde{I}(\tilde{r}) = -\nabla_r \cdot \tilde{H}(\tilde{r}) \ . \tag{60}$$

Eqs. (59) and (60) follow from the fact that the hydrodynamic variables are conserved. Substitution of Eqs. (59) and (60) into Eq. (53) yields:

$$\dot{\tilde{q}}(\tilde{r},t) = -\nabla_r \cdot \langle\tilde{J}(\tilde{r})\rangle_L(t) + \nabla_r \cdot \int_0^\infty \langle\tilde{H}(\tilde{r},\tau)\tilde{H}(\tilde{r}')\rangle_L(t)$$

$$* \ \nabla_{r'}\beta\Phi(\tilde{r}',t)d\tau \tag{61}$$

We now use the fact that the $q(r,t)$ and $\Phi(r,t)$ are slowly varying in space whereas the correlation functions $\langle\tilde{H}(r,\tau)\tilde{H}(r')\rangle_L$ decay quickly to zero as $|r - \tilde{r}'|$ increases. We define the transport coefficients, $\underset{\approx}{\chi}(\tilde{r},t)$, by

$$\underset{\approx}{\chi}(\tilde{r},t) \equiv \int_0^\infty d\tau \int_V dr' \langle\tilde{H}(\tilde{r},\tau)\tilde{H}(\tilde{r}')\rangle_L(t) \ . \tag{62}$$

Substitution into Eq. (61) yields the final general form for the hydrodynamic equations:

$$\dot{q}(r,t) = -\nabla_{\Omega} \cdot \langle J(r)\rangle_L(t) + \nabla_r \cdot \chi(r,t) \cdot \nabla_r \beta\Phi(r,t) . \tag{63}$$

Note that the space and time dependence of $\chi(r,t)$ arises completely from the $\Phi(r,t)$ which occur in Eq. (51). Thus, the χ's are slowly varying in space and time and can be written as $\chi[\Phi(r,t)]$.

The linearized form of Eq. (63) is:

$$\dot{q}(r,t) = -\beta\int_V dr'\langle J(r)Q(r')\rangle \cdot \nabla_r \Phi(r,t)$$

$$+ \nabla_r \cdot \chi \cdot \nabla_r \beta\Phi(r,t) , \tag{64}$$

where

$$\chi = \int_0^\infty d\tau \int_V dr'\langle H(r,\tau)H(r')\rangle \tag{65}$$

is independent of r and t and Φ is defined by

$$q(r,t) \equiv \beta K(r,r';0) * \Phi(r',t) . \tag{66}$$

The explicit forms of Eq. (63) are given in reference 6.

CONCLUSIONS

We have seen that some care must be taken in the use of response theory. Even with adiabatic forces of the form of Eq. (6), the distribution function at time, t = 0 , is not of the local equilibrium form. Indeed, the expression for $q(r,t)$ in terms of $F(r)$, Eq. (25), or its inverse, is not of the form that can be obtained from a local equilibrium function. However, it is possible to define a time dependent local equilibrium distribution function, Eq. (51), which yields the correct results for the transport equations, Eqs. (41) and (63).

REFERENCES

1a. J.L. Jackson, Phys. Rev. 87, 471 (1952).
1b. H.B. Callen, M.L. Barasch, J.L. Jackson, Phys. Rev. 88, 1382-86 (1952).
1c. J.L. Jackson, P. Mazur, Physica 30, 2295-2304 (1964).
2. J.L. Jackson, private communication.
3. N.G. van Kampen, Physica Norwegica 5, 10 (1971).
4. See e.g. T. Keyes and I. Oppenheim, Phys. Rev. A7, 1384 (1973).
5. P. Mazur and I. Oppenheim, Physica 23, 197 (1957).
6. J. Weare and I. Oppenheim, Physica 72, 20 (1974).

Dr. Oppenheim is Professor of Chemistry at MIT. He was Chief of Theoretical Physics at General Dynamics/Convair and was "van der Waals Professor" at the University of Amsterdam. He is a fellow of the American Physical Society and of the American Academy of Arts and Sciences. Recently his concern has been in generalized fluid dynamics. He met Julius in Washington, D.C. and they worked together at the National Bureau of Standards.

APPLICATION OF NONLINEAR STABILITY THEORY TO THE STUDY OF THE EFFECTS OF DIFFUSION ON PREDATOR-PREY INTERACTIONS

Lee A. Segel
Department of Applied Mathematics, Weizmann Institute of Science, Rehovot, Israel, and Rensselaer Polytechnic Institute, Troy, New York 12181

Simon A. Levin
Section of Ecology and Systematics, and Department of Theoretical and Applied Mechanics, Cornell University, Ithaca, New York 14853

ABSTRACT

In the basic predator-prey model under investigation, growth rates of both species are influenced by both predator and prey population levels. Random dispersal is modelled by diffusion-like terms in both discrete and continuous situations. Application of a combined successive-approximations multiple-scale approach to stability theory shows that under some circumstances uniform conditions will be succeeded by a new steady state wherein predator and prey are more concentrated in certain regions. Under other circumstances (in the discrete case only), although uniform conditions are stable according to linear theory, nevertheless sufficiently large perturbations are destabilizing. The biological implications of the results are discussed.

PREFACE

In a now famous paper, Turing[1] showed that spatial patterns can arise from a combination of chemical reaction and diffusion. Turing suggested that such pattern formation could well be of major importance in developmental biology. Patterns of finite wavelength were only obtained in situations wherein chemical equilibrium was stable in the absence of diffusion. Thus, diffusion, normally regarded as an influence that tends to erase zones of relatively high chemical concentration, can act to promote instability and inhomogeneity.

In spite of various contributions to the theory by a number of authors, a simple physical explanation for "diffusive instability" was lacking. The situation was remedied by Segel and Jackson[2] who showed how diffusion can cause too rapid decay of the chemical that stabilizes an interaction, thereby permitting a continual increase of the local concentration of the destabilizer and a consequent instability. Segel and Jackson[2] then went on to show how such diffusive instabilities can arise from the effects of random dispersal on models of predator-prey interactions, and Levin[3] independently treated the same problem for discrete intercommunicating "patches" of species. This is a natural step, for both the morphogenetic and ecological models have the same mathematical structure of diffusion equations augmented by nonlinear reaction terms.

The present study extends the ecological part of Segel and Jackson[2] and the work of Levin[3] by taking nonlinear effects into account. Moreover, much more attention is given to the ecological background of the analysis.

This work is dedicated to the memory of Julius Jackson, who we trust would have approved. Indeed, the authors plus Julius (posthumously) and Ralph Mitchell of Harvard recently received a joint grant from the U.S.-Israel Binational Fund to continue research of this general nature. On turning now to the details of our investigation and its background, we are acutely aware of the loss we have suffered from Julius' untimely passing.

INTRODUCTION

In most natural communities, it is not possible to isolate predator-prey interactions and view density changes in these two species as mutually causative and independent of other species. Predators often possess alternative prey, prey are affected by interspecific competitors, and the role of predation is to modify competitive interactions rather than to account entirely for prey limitation. Typical in demonstrating this effect are the experiments of Paine[4,5] in intertidal communities, Porter[6] on coral reefs, Harper[7] in vegetational systems, and Brooks and Dodson,[8] Zaret,[9] and Zaret and Paine[10] in lakes. Moreover, Pimentel[11] has argued that evolutionary processes will tend to minimize the importance of the classical predator-prey linked density fluctuations.

Nevertheless, examples of strongly associated predator and prey fluctuations are known, although in most cases the evidence is correlative and no clear mutual causative process is identifiable. Early mathematical models of Lotka[12] and Volterra[13,14] predicting oscillations in predator-prey systems stimulated laboratory experiments on protozoans by Gause[15] and more recently by Luckinbill.[16] Huffaker,[17] in a classic study on predation using two species of mites, demonstrated the importance of differential dispersal abilities in maintaining sustained oscillations in his experimental system.

Gause pointed out that the role of experiments such as his is at the very least to isolate for examination "elementary interactions." Mathematical predator-prey models can also play such a role; they abstract and extend the role of the laboratory.

Our discussion will be based upon a continuous time model consisting of a system of two differential equations relating change in density of the predator population E (for exploiter, notation following Rosenzweig[18]) to that of the prey V (for victim). Letting $P(E,V)$ equal the predation rate and ζ the conversion efficiency, one obtains

$$dV/dt = V\,f(V) - P(E,V), \quad dE/dt = E\,g(E) + \zeta P(E,V)\ . \qquad (1)$$

Here, $f(V)$ represents the per capita growth rate of the prey in the absence of predation and $g(E)$ the growth rate (generally negative) of predator in the absence of prey. This general framework, and especially the assumption that ζ is constant, ignores change in genetics and age structure. In some situations, such an assumption may not be justified.[11,19] Temporal variation in the functional forms is also ignored.

The most usual assumption is that $P(E,V) = E\phi(V)$.[20] In the Lotka-Volterra models, $\phi(V)$ is a linear and increasing function of V; but a more reasonable form incorporates the notion of the functional response of the prey[21] and treats ϕ as an increasing, but uniformly bounded, function of V. Of particular interest is the "Type 2" functional response justified empirically and theoretically by Holling.[21] Here predator handling time of prey is considered, and $\phi(V)$ assumes the form

$$\phi(V) = aV/(1+bV),$$

where a and b are positive constants.

Since $d[V^{-1}\phi(V)]/dV < 0$, an autocatalytic influence is present in the equation for V in (1). The autocatalytic term tends to disappear as V increases, since $\phi(V) \to a/b$; but it may be important for V small and may outweigh whatever damping is present in f. In short, when prey are scarce, there may be safety in numbers represented by an autocatalytic effect of V on per capita growth of V.

This can be cast in a more general framework. The results we shall derive do not depend directly on the form (1), but in fact we may begin from the much more general description

$$dV/dt = VF(V,E), \quad dE/dt = EG(V,E). \tag{2}$$

We assume here only that $\partial F/\partial E < 0$ and $\partial G/\partial V > 0$; that is, predators are bad for individual prey, but prey are good for predators. Where the prey are self-damped, $\partial F/\partial V < 0$; but where the autocatalytic effect obtains, $\partial F/\partial V > 0$. If the predators are self-damped, $\partial G/\partial E < 0$; but the possibility of an autocatalytic term ($\partial G/\partial E > 0$) exists there as well. The autocatalytic term, analogous in effect to what is called in the ecological literature an _Allee effect_,[22] may arise for a variety of reasons including, for example, ease of finding mates or the functional response described above.

THE HOMOGENEOUS CASE - STABILITY OF EQUILIBRIUM

Interest will focus henceforth on the system (2). This system may possess a variety of equilibria [points $\hat{V}$, $\hat{E}$ for which $F(\hat{V},\hat{E}) = G(\hat{V},\hat{E}) = 0$]. We assume an equilibrium exists, and examine the system in a neighborhood of that point. (At first we recapitulate in terse form the results of Segel and Jackson[2] and Levin.[3]) For that purpose, it is sufficient to replace F and G by linear approximations, and (2) is thereby replaced by the system

$$dV/dt = V(K - \alpha V - \beta E), \quad dE/dt = E(-L + \gamma V - \delta E). \tag{3}$$

Here K, L, β, and γ are assumed positive, but we shall have more to say about α and δ. Note here that if either α or δ is negative, an Allee effect is present.

The equilibrium in the system (3) occurs at

$$\hat{V} = (K\delta + L\beta)/(\alpha\delta + \beta\gamma), \quad \hat{E} = (K\gamma - L\alpha)/(\alpha\delta + \beta\gamma). \tag{4}$$

By assumption, $\hat{V} > 0$ and $\hat{E} > 0$. Moreover, the linearization of (3) at the equilibrium (4) is seen to hinge on the matrix

$$\underset{\sim}{M} = \begin{pmatrix} -\alpha\hat{V} & -\beta\hat{V} \\ \gamma\hat{E} & -\delta\hat{E} \end{pmatrix}. \tag{5}$$

Linear stability of the equilibrium (4) occurs if and only if all eigenvalues of the matrix $\underset{\sim}{M}$ have negative real parts (we ignore the details involving the borderline case of zero real part). This requires, in particular, that

$$(\det \underset{\sim}{M})/\hat{V}\hat{E} = \alpha\delta + \beta\gamma > 0. \tag{6}$$

Substitution of this requirement into (4) and utilization of the conditions $\hat{V} > 0$ and $\hat{E} > 0$ leads to the further conditions

$$K\delta + L\beta > 0 \tag{7a}$$

and

$$K\gamma - L\beta > 0. \tag{7b}$$

Finally, stability requires

$$\mathrm{Tr}\ M = -\alpha\hat{V} - \delta\hat{E} < 0;$$

that is,

$$-\alpha(K\delta + L\beta) - \delta(K\gamma - L\alpha) < 0. \tag{8}$$

These conditions [(6)-(8)] are met by a wide variety of parameters. They require, in particular, that either α or δ be positive; that is, an Allee effect in both species will always result in instability. On the other hand, they tolerate an Allee effect in one species, provided the self-damping of the other predominates.

It is not of interest here to catalogue the cases for which (6)-(8) are satisfied. Suffice it to say, such cases exist; and the exact regions can be obtained without undue effort. Even if the equilibria are unstable, the systems considered are of great interest. Indeed, May,[23] using results of Kolmogorov,[24] has shown that most of the unstable, nonlinear systems proposed for predator-prey interactions possess stable limit cycles.

SCOPE OF THE PRESENT WORK

It will be our purpose to examine the effects of the introduction of dispersion and dispersal terms upon the stability of the equilibrium points. This complements the work of Maynard Smith,[20] who has discussed in a qualitative way the effects of dispersion and dispersal when limit cycles are present, utilizing the general approach of Winfree.[25]

Interest in dispersal and dispersion patterns in systems of interacting populations traces back at least to Hutchinson[26] and Skellam.[27] Huffaker[17] found that the persistence of laboratory predator-prey systems was critically dependent upon the differential dispersal abilities of his organisms. One can demonstrate the potential persistence of systems in which the prey is fugitive by equations which consider only the fractions of patches occupied, allowing for local extinctions of prey populations (see, for example, Van der Meer[28]). Such approaches, however, do not consider either transient dynamics or spatial patterns.

Using a different approach, we focus on two types of environments, discrete (patchy) and continuous, and develop analogous results for each. The approach to discrete environments follows that developed in Levin,[3] while an introduction to the continuous theory is provided in Segel and Jackson.[2] We shall assume a homogeneous environment with regard to habitat conditions, so that the reaction kinetics as given by (3) provide the description locally at every point in space.

By means of linear stability theory, we first review the possible destabilizing effects of random dispersal. Concentrating for simplicity on a somewhat specialized model, we then turn to the effects of non-linearity. Application of a combined successive-approximations multiple-scale procedure shows that in the continuous case the unstable homogeneous state transforms in time to a steady non-homogeneous distribution of organisms. The same can happen in the discrete case, but for certain parameter ranges new phenomena can appear (see below).

INCLUDING THE EFFECTS OF DISPERSION AND DISPERSAL

The discrete model assumes that the environment consists of a set of m patches, numbered 1, . . ., m, with prey and predator densities in individual patches represented by variables V_i, E_i. The equations representing change in V_i, E_i and incorporating movement take the form

$$dV_i/dt = V_i(K - \alpha V_i - \beta E_i) + \sum_j d_{ij}(V_j - V_i),$$
$$dE_i/dt = E_i(-L + \gamma V_i - \delta E_i) + \sum_j d'_{ij}(E_j - E_i). \qquad (9)$$

In this formulation, movement is by passive dispersal akin to diffusion. More complicated movement patterns, in particular ones incorporating predator response to prey gradients, can be substituted, but would obscure the results we seek here to illuminate. The quadratic forms of the reaction kinetics are no restriction, since we shall always be near equilibrium.

We assume $d_{ij} = c_{ij}\tilde{\mu}$, $d'_{ij} = c_{ij}\tilde{\nu}$ and $c_{ij} = c_{ji}$. Thus the relative diffusion rates of predator and prey between any pair of patches are proportional to a symmetric connectivity measure (c_{ij}). Hence d_{ij} and d'_{ij} always bear the same fixed proportional relationship to one another, $d_{ij}/d'_{ij} = \tilde{\mu}/\tilde{\nu}$; that is, the ratio is the same for any pair of patches and measures the relative dispersal abilities of the two species.

The continuous version of (9) replaces V_i, E_i by functions $V(t,\underset{\sim}{x})$, $E(t,\underset{\sim}{x})$ of space and time, and results in equations

$$\begin{aligned} \partial V/\partial t &= V(K - \alpha V - \beta E) + \nabla\cdot(\mu\nabla V), \\ \partial E/\partial t &= E(-L + \gamma V - \delta E) + \nabla\cdot(\nu\nabla E), \end{aligned} \tag{10}$$

where in general, μ and ν can vary as functions of $\underset{\sim}{x}$. We shall, however, take μ and ν to be constants.

Both systems (9) and (10) permit a uniform spatial equilibrium, given by $E = \hat{E}$, $V = \hat{V}$, where $\hat{E}$ and $\hat{V}$ are as in (4). We shall generally assume the conditions (6)-(8) are satisfied, so that the equilibrium exists and is stable. If this is not done, as we shall see, dispersal cannot stabilize the situation. When the effects of dispersal are ignored, then $d_{ij} \equiv d'_{ij} \equiv 0$ in (9), and $\tilde{\mu} \equiv \tilde{\nu} \equiv 0$ in (10).

STABILITY IN THE DISCRETE CASE

The full linear stability analysis of the system (9) can be shown to depend upon the eigenvalues of the matrix

$$\underset{\sim}{M} = \underset{\sim}{I}_n \otimes \begin{pmatrix} -\alpha\hat{V} & -\beta\hat{V} \\ \gamma\hat{E} & -\delta\hat{E} \end{pmatrix} + \underset{\sim}{\Delta} \otimes \begin{pmatrix} \tilde{\mu} & 0 \\ 0 & \tilde{\nu} \end{pmatrix}$$

where $\otimes$ denotes tensor product, $\underset{\sim}{I}_n$ is the n-dimensional identity matrix, and the "structural" matrix $\underset{\sim}{\Delta}$ has components Δ_{ij}. Here $\Delta_{ij} = c_{ij}$, $i \neq j$; $\Delta_{ii} = -\sum_{j\neq i} c_{ij}$. For a full discussion of this method of analysis, see Othmer and Scriven.[29,30] In particular, it follows[29,31] that the eigenvalues of M are the eigenvalues of the matrices

$$\underset{\sim}{M}_\lambda = \begin{pmatrix} -\alpha\hat{V} + \lambda\tilde{\mu} & -\beta\hat{V} \\ \gamma\hat{E} & -\delta\hat{E} + \lambda\tilde{\nu} \end{pmatrix} \tag{11}$$

where λ is an eigenvalue of the structural matrix. Now recall that $\Delta_{ii} = -\sum_{j\neq i} \Delta_{ij}$ for each i. Thus the columns of $\underset{\sim}{\Delta}$ are linearly dependent and so $\underset{\sim}{\Delta}$ is singular. Therefore, at least one eigenvalue of $\underset{\sim}{\Delta}$ vanishes, so at least one pair of eigenvalues of $\underset{\sim}{M}$ does not change with $\tilde{\mu}$ and $\tilde{\nu}$. Consequently, dispersal of the sort proposed could never stabilize an unstable interaction.

For the specific assumptions we have made concerning dispersal, $\underset{\sim}{\Delta}$ is real and symmetric; hence, its eigenvalues are real. As we have already remarked, $\underset{\sim}{\Delta}$ is singular; and by Gershgôrin's theorem[32] every eigenvalue is contained in a disc $|\lambda - \Delta_{ii}| \leq \sum_{j \neq i} |\Delta_{ij}|$ for some i. It follows that $\underset{\sim}{\Delta}$ has no positive eigenvalues since all row sums of $\underset{\sim}{\Delta}$ are zero and $\Delta_{ij} \geq 0$ for $i \neq j$. Thus $\lambda \leq 0$ in (11), and the trace of $\underset{\sim}{M}_\lambda$ is negative provided the equilibrium was stable without dispersal. Instability can still set in if for some λ the determinant is negative, i.e. if

$$(\lambda\tilde{\mu} - \alpha\hat{V})(\lambda\tilde{\nu} - \delta\hat{E}) + \beta\gamma\hat{E}\hat{V} < 0. \tag{13}$$

Since each $\lambda \leq 0$, it is not difficult to see that this can only happen provided $\alpha\delta < 0$; that is, provided there is an Allee effect present in one species.

On the other hand, if the Allee effect is present, the system will destabilize (as $\tilde{\mu},\tilde{\nu}$ increase) at critical values of $\tilde{\mu}$ and $\tilde{\nu}$. For example, if $\alpha < 0 < \delta$ and $\tilde{\mu} = 0$, destabilization will occur at the point $\tilde{\nu} = (\alpha\delta + \beta\gamma)E/(\alpha\lambda^*)$, where λ^* is the dominant eigenvalue of the structural matrix (the eigenvalue of largest magnitude). For other values of $\tilde{\mu}$, the exact boundary of destabilization (Figure 1) will depend upon the other eigenvalues of the structural matrix, and the pattern for a highly reticulated system tends to that shown in Figure 2. (For details, see Appendix 1.)

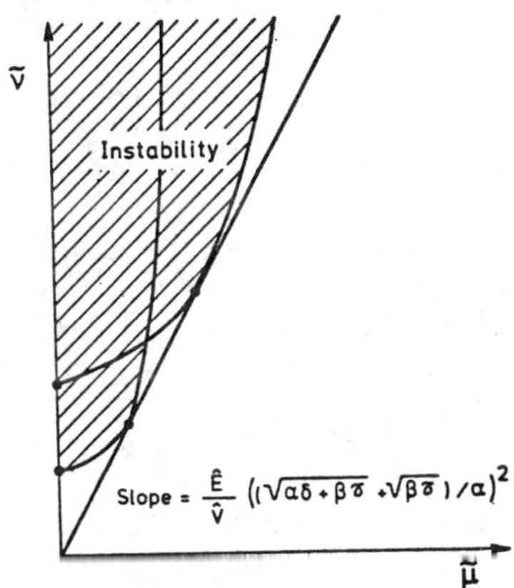

Figure 1. Stability diagram in the discrete case.

For n = 2, the structural matrix has a single non-zero eigenvalue, and the boundary of the region of instability will be without bumps. As n becomes large, the boundary will in general smooth out and approach the straight line that bounds the region of instability in the continuous case.

STABILITY IN THE CONTINUOUS CASE

In the continuous case (10), the results are similar. Stability is evaluated with respect to perturbations of the form

$$(v, e)^T = (V - \hat{V}, E - \hat{E})^T = \underset{\sim}{C} \exp(\sigma t) \exp(i\underset{\sim}{k}\cdot\underset{\sim}{x})$$

where σ is the constant growth rate of the perturbation, $\underset{\sim}{k}$ is a constant wave number vector, and $\underset{\sim}{C}$ is another constant vector. Setting $q = |\underset{\sim}{k}|$, we find that stability depends on the stability properties of the matrices

$$\underset{\sim}{M}_q = \begin{pmatrix} -\alpha\hat{V} - \mu q^2 & -\beta\hat{V} \\ \gamma\hat{E} & -\delta\hat{E} - \nu q^2 \end{pmatrix} \tag{14}$$

in analogy to (11). Note however, that μ and ν do not have exactly the same interpretation as $\tilde{\mu}$ and $\tilde{\nu}$ of the discrete model. The diffusivities μ and ν are purely local, while $\tilde{\mu}$ and $\tilde{\nu}$ must include in general a factor which is sensitive to the distances between patches.

In analogy to (13), it is seen that the equilibrium is unstable to disturbances of wave number q if and only if

$$(-\mu q^2 - \alpha\hat{V})(-\nu q^2 - \delta\hat{E}) + \beta\gamma\hat{E}\hat{V} < 0. \tag{15}$$

Inequality (15) holds for a range of positive q^2 (i.e. the equilibrium is unstable) whenever

$$\alpha\hat{V}\nu + \delta\hat{E}\mu < -2\sqrt{(\alpha\delta + \beta\gamma)\mu\nu\hat{E}\hat{V}}.$$

One can transform this condition easily to the condition

$$\hat{V}\nu > \hat{E}\mu[(\sqrt{\alpha\delta + \beta\gamma} + \sqrt{\beta\gamma})/\alpha]^2.$$

This region is shown in Figure 2. It is essentially the limiting case of Figure 1. Although ν and μ do not have the same interpretation as $\tilde{\nu}$ and $\tilde{\mu}$, ν/μ does have the same interpretation as $\tilde{\nu}/\tilde{\mu}$.

Provided $\mu \neq 0$, it is convenient to introduce

$$\phi^2 = \hat{V}\nu/\hat{E}\mu.$$

For a given wave number $q > 0$, there exists a value $\phi_c(q;\mu)$ such that when $\phi > \phi_c(q;\mu)$, destabilization occurs. Assuming that all wave numbers are available (unbounded region), instability is bound to occur at some wavelength when

$$\phi > \phi_c = \min_q \phi_c(q;\mu) = (\sqrt{\alpha\delta + \beta\gamma} + \sqrt{\beta\gamma})/|\alpha|.$$

Note that ϕ_c is independent of μ.

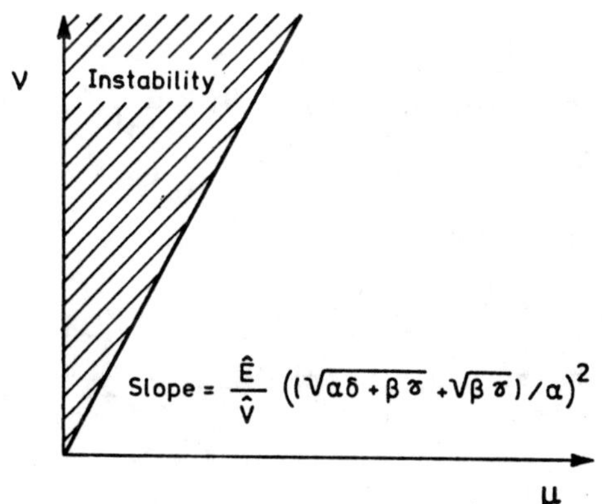

Figure 2. Stability diagram in the continuous case.

SPECIALIZATION AND NONDIMENSIONALIZATION

We now elect to restrict our example somewhat, for purposes of clarity. All the results to follow carry over to the general case, but maintenance of the full generality would unnecessarily complicate the situation.

To begin, we have seen that the spatially uniform equilibrium, if initially stable, remains stable under the influence of dispersal unless an Allee effect is present in one species. This result is true in both the continuous and discrete cases. We focus interest here on the cases which do destabilize, and make the specific assumption that the Allee effect is in the prey. The analogous results for an Allee effect in the predator carry over immediately.

We make the further simplification L = 0 in (3). Since $\alpha < 0 < \delta$, this essentially only changes the actual values of the equilibrium as given by (4), but simplifies later calculations immensely.

It is convenient to introduce dimensionless variables according to the formulae

$$\bar{e} = E\ \delta/K, \quad \bar{v} = V\ \gamma/K, \quad \bar{t} = Kt. \tag{16}$$

In the continuous case, provided $\mu \neq 0$,

$$\bar{x} = x/(\mu/K)^{-\frac{1}{2}}, \quad \bar{y} = y/(\mu/K)^{-\frac{1}{2}}, \quad \theta^2 = \nu/\mu. \tag{17}$$

In the discrete case, provided $\tilde{\mu} \neq 0$,

$$\theta^2 = \tilde{\nu}/\tilde{\mu}. \tag{18}$$

This transforms the basic equations into

$$\partial\bar{v}/\partial\bar{t} = (1 + \bar{k}\bar{v})\bar{v} - a\bar{e}\bar{v} + \nabla^2\bar{v}, \quad \partial\bar{e}/\partial\bar{t} = \bar{e}\bar{v} - \bar{e}^2 + \theta^2\nabla^2\bar{e} \tag{19}$$

in the continuous case and

$$d\bar{v}_i/d\bar{t} = (1 + \bar{k}\bar{v}_i)\bar{v}_i - a\bar{e}_i\bar{v}_i + \Sigma\, \bar{d}_{ij}(\bar{v}_j - \bar{v}_i)$$
$$d\bar{e}_i/d\bar{t} = \bar{e}_i\bar{v}_i - \bar{e}_i^{\,2} + \theta^2\, \Sigma\, \bar{d}_{ij}(\bar{e}_j - \bar{e}_i) \tag{20}$$

in the discrete. Here

$$\bar{k} = -\alpha/\gamma, \quad a = \beta/\delta, \quad \bar{d}_{ij} = d_{ij}/K, \text{ and } \bar{d}'_{ij} = d'_{ij}/K. \tag{21}$$

Both sets of equations have a non-trivial <u>steady-state</u> <u>spatially</u> <u>uniform</u> <u>solution</u>. This is given by

$$\bar{e} \equiv \bar{v} \equiv p^{-2} \tag{22}$$

in the continuous case, and

$$\bar{e}_i \equiv \bar{v}_i \equiv p^{-2} \tag{23}$$

in the discrete. Here,

$$p^2 = a - \bar{k}. \tag{24}$$

We assume that $a - \bar{k} > 0$, corresponding to the condition $\alpha\delta + \beta\gamma > 0$ which is a prerequisite for stability without dispersal [condition (6)].

We employ as dependent variables the deviations e and v from steady-state densities where

$$e = \bar{e} - p^{-2}, \quad v = \bar{v} - p^{-2}. \tag{25}$$

The equations now take the form

$$-(\partial \underset{\sim}{V}/\partial\bar{t}) + \underset{\sim}{M}_\theta \underset{\sim}{V} = \underset{\sim}{N}(\underset{\sim}{V}) \tag{26}$$

in the continuous case. Here

$$\underset{\sim}{V} = (v, e)^T, \quad \underset{\sim}{M}_\theta = \begin{pmatrix} \bar{k}p^{-2} + \nabla^2 & -ap^{-2} \\ p^{-2} & -p^{-2} + \theta^2\nabla^2 \end{pmatrix},$$
$$\underset{\sim}{N}(\underset{\sim}{V}) = (aev - \bar{k}v^2,\ e^2 - ev)^T. \tag{27}$$

In the discrete case, one may write

$$-(d\underset{\sim}{V}/d\bar{t}) + \underset{\sim}{M}^*_\theta \underset{\sim}{V} = \underset{\sim}{N}(\underset{\sim}{V}) \tag{28}$$

where

in the continuous case; and

$$\begin{aligned} d\bar{v}_i/d\bar{t} &= (1 + \bar{k}\bar{v}_i)\bar{v}_i - a\bar{e}_i\bar{v}_i + \Sigma\, d_{ij}(\bar{v}_j - \bar{v}_i) \\ d\bar{e}_i/d\bar{t} &= \bar{e}_i\bar{v}_i - \bar{e}_i{}^2 + \theta^2\, \Sigma\, \bar{d}_{ij}(\bar{e}_j - \bar{e}_i) \end{aligned} \tag{20}$$

in the discrete. Here

$$\bar{k} = -\alpha/\gamma, \quad a = \beta/\delta, \quad \bar{d}_{ij} = d_{ij}/K, \text{ and } \bar{d}'_{ij} = d'_{ij}/K. \tag{21}$$

Both sets of equations have a non-trivial <u>steady-state</u> <u>spatially</u> <u>uniform</u> <u>solution</u>. This is given by

$$\bar{e} \equiv \bar{v} \equiv p^{-2} \tag{22}$$

in the continuous case, and

$$\bar{e}_i \equiv \bar{v}_i \equiv p^{-2} \tag{23}$$

in the discrete. Here,

$$p^2 = a - \bar{k}. \tag{24}$$

We assume that $a - \bar{k} > 0$, corresponding to the condition $\alpha\delta + \beta\gamma > 0$ which is a prerequisite for stability without dispersal [condition (6)].

We employ as dependent variables the deviations from steady-state densities e and v, where

$$e = \bar{e} - p^{-2}, \quad v = \bar{v} - p^{-2}. \tag{25}$$

The equations now take the form

$$-(\partial \underset{\sim}{V}/\partial\bar{t}) + \underset{\sim}{M}_\theta \underset{\sim}{V} = \underset{\sim}{N}(\underset{\sim}{V}) \tag{26}$$

in the continuous case. Here

$$\begin{aligned} \underset{\sim}{V} &= (v, e)^T, \quad \underset{\sim}{M}_\theta = \begin{pmatrix} \bar{k}p^{-2} + \nabla^2 & -ap^{-2} \\ p^{-2} & -p^{-2} + \theta^2\nabla^2 \end{pmatrix}, \\ \underset{\sim}{N}(\underset{\sim}{V}) &= (aev - \bar{k}v^2,\ e^2 - ev)^T. \end{aligned} \tag{27}$$

In the discrete case, one may write

$$-(d\underset{\sim}{V}/d\bar{t}) + \underset{\sim}{M}^*_\theta \underset{\sim}{V} = \underset{\sim}{N}(\underset{\sim}{V}) \tag{28}$$

where

$$\underset{\sim}{V} = (v_1, e_1, v_2, e_2, \ldots, v_n, e_n)^T, \quad \underset{\sim}{M}^*_\theta = \underset{\sim}{I}_n \otimes \underset{\sim}{W} + \tilde{\mu}\underset{\sim}{\bar{\Delta}} \otimes \begin{pmatrix} 1 & 0 \\ 0 & \theta^2 \end{pmatrix},$$

$$\underset{\sim}{W} = p^{-2} \begin{pmatrix} \bar{k} & -a \\ 1 & -1 \end{pmatrix}, \quad \underset{\sim}{\bar{\Delta}} = \underset{\sim}{\Delta}/K, \tag{29}$$

and

$$N(\underset{\sim}{V}) = (ae_1v_1 - \bar{k}v_1^2, e_1^2 - e_1v_1, ae_2v_2 - \bar{k}v_2^2, e_2^2 - e_2v_2, \ldots,$$
$$ae_nv_n - \bar{k}v_n^2, e_n^2 - e_nv_n)^T.$$

Following the procedures outlined in the last section, one finds that in the continuous case, the equilibrium is stable with respect to a perturbation

$$\underset{\sim}{V} = \underset{\sim}{C} \exp(\sigma t) \exp(i\underset{\sim}{k} \cdot \underset{\sim}{x}) \tag{30}$$

provided

$$\theta^2 q^4 + p^{-2}(1 - \bar{k}\theta^2)\, q^2 + p^{-2} > 0. \tag{31}$$

Here, again, $q = |\underset{\sim}{k}|$. Recall that $p = (a - \bar{k})^{\frac{1}{2}}$. Further, we introduce the new variable $m = a^{\frac{1}{2}}$. In terms of these new parameters, the original parameters are given by

$$a = m^2, \quad \bar{k} = m^2 - p^2. \tag{32}$$

We have assumed that $\bar{k} > 0$, since $\alpha < 0 < \gamma$, and so $m > p$. Further, we assume $\bar{k} < 1$ [the equivalent of (8)] to guarantee stability in the absence of dispersal. Thus, our assumptions are

$$0 < \bar{k} < 1; \quad \text{i.e., } 0 < m^2 - p^2 < 1. \tag{33}$$

The region (31) has as boundary the curve

$$\theta^2 q^4 + p^{-2}(1 - \bar{k}\theta^2)\, q^2 + p^{-2} = 0 \tag{33a}$$

which is graphed in Figure 3. The minimum for this curve occurs at $(\theta, q) = (\theta_c, q_c)$, where

$$\theta_c = (m-p)^{-1}, \quad q_c^2 = (m-p)/p. \tag{34}$$

Thus θ_c^2 is the <u>critical diffusivity ratio</u>, in that the steady state solution is stable to small perturbations when $\theta < \theta_c$ but unstable when $\theta > \theta_c$.

If θ is increased slowly for some reason to beyond θ_c, instability would be expected to set in for a perturbation having the critical wave number. Allowing for our slight changes in notation,

$$\underset{\sim}{V} = (v_1, e_1, v_2, e_2, \ldots, v_n, e_n)^T, \ \underset{\sim}{M}^*_\theta = \underset{\sim}{I}_n \otimes \underset{\sim}{W} + \mu\underset{\sim}{\bar{\Delta}} \otimes \begin{pmatrix} 1 & 0 \\ 0 & \theta^2 \end{pmatrix},$$

$$\underset{\sim}{W} = p^{-2} \begin{pmatrix} \bar{k} & -a \\ 1 & -1 \end{pmatrix}, \quad \underset{\sim}{\bar{\Delta}} = \underset{\sim}{\Delta}/K, \tag{29}$$

and

$$\underset{\sim}{N}(\underset{\sim}{V}) = (ae_1v_1 - \bar{k}v_1^2, \ e_1^2 - e_1v_1, \ ae_2v_2 - \bar{k}v_2^2, \ e_2^2 - e_2v_2, \ \ldots,$$
$$ae_nv_n - \bar{k}v_n^2, \ e_n^2 - e_nv_n)^T.$$

Following the procedures outlined in the last section, one finds that in the continuous case, the equilibrium is stable with respect to a perturbation

$$\underset{\sim}{V} = \underset{\sim}{C} \exp(\sigma t) \exp(i\underset{\sim}{k} \cdot \underset{\sim}{x}) \tag{30}$$

provided

$$\theta^2 q^4 + p^{-2}(1 - \bar{k}\theta^2) q^2 + p^{-2} > 0. \tag{31}$$

Here, again, $q = |\underset{\sim}{k}|$. Recall that $p = (a - \bar{k})^{1/2}$. Further, we introduce the new variable $m = a^{1/2}$. In terms of these new parameters, the original parameters are given by

$$a = m^2, \quad \bar{k} = m^2 - p^2. \tag{32}$$

We have assumed that $\bar{k} > 0$, since $\alpha < 0 < \gamma$, and so $m > p$. Further, we assume $\bar{k} < 1$ [the equivalent of (8)] to guarantee stability in the absence of dispersal. Thus, our assumptions are

$$0 < \bar{k} < 1; \quad \text{i.e.,} \ 0 < m^2 - p^2 < 1. \tag{33}$$

The region (31) has as boundary the curve

$$\theta^2 q^4 + p^{-2}(1 - \bar{k}\theta^2) q^2 + p^{-2} = 0 \tag{33a}$$

which is graphed in Figure 3. The minimum for this curve occurs at $(\theta, q) = (\theta_c, q_c)$, where

$$\theta_c = (m-p)^{-1}, \quad q_c^2 = (m-p)/p. \tag{34}$$

Thus θ_c^2 is the <u>critical diffusivity ratio</u>, in that the steady state solution is stable to small perturbations when $\theta < \theta_c$ but unstable when $\theta > \theta_c$.

If θ is increased slowly for some reason to beyond θ_c, instability would be expected to set in for a perturbation having the critical wave number. Allowing for our slight changes in notation, the results here for θ_c and q_c agree with those of Segel and Jackson.[2]

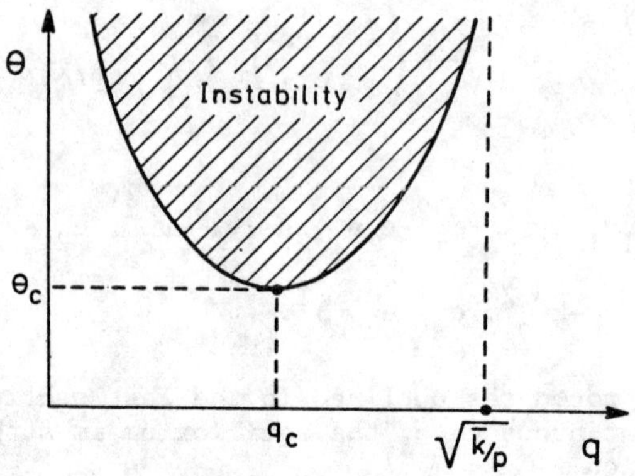

Figure 3. Stability diagram in the continuous case.

Further results of linear theory will prove useful. (30) will be a solution of the linear equation

$$-(\partial \underset{\sim}{V}/\partial \bar{t}) + \underset{\sim}{M}_\theta \underset{\sim}{V} = 0$$

if and only if $\underset{\sim}{M}_{\theta,q}\underset{\sim}{C} = \sigma_{\theta,q}\underset{\sim}{C}$, where

$$\underset{\sim}{M}_{\theta,q} = \begin{pmatrix} \bar{k}p^{-2} - q^2 & -ap^{-2} \\ p^{-2} & -p^{-2} - \theta^2 q^2 \end{pmatrix}.$$

Let us denote the two eigenvalues by

$$\sigma^{(i)}_{\theta,q}, \quad i = 1,2; \quad \sigma^{(1)}_{\theta,q} \geq \sigma^{(2)}_{\theta,q}; \tag{35}$$

when $\theta \approx \theta_c$, we have

$$\begin{aligned} \sigma^{(1)}_{\theta,q_c} &= [2(m-p)^2(\theta-\theta_c)]/[p\{1-(m-p)^2\}] + O(\theta-\theta_c)^2, \\ \sigma^{(2)}_{\theta,q_c} &= -q_c^{\;2}(1+\theta_c^{\;2}) - p^{-2}(1-m^2+p^2) + O(\theta-\theta_c). \end{aligned} \tag{36}$$

The corresponding eigenvectors will be denoted by $\underset{\sim}{C}_1$ and $\underset{\sim}{K}_1$ respectively, where

$$\underset{\sim}{C}_1 = (m\theta_c, 1)^T + O(\theta-\theta_c), \quad \underset{\sim}{K}_1 = (m/\theta_c, 1)^T + O(\theta-\theta_c). \tag{37a}$$

For $\theta \approx \theta_c$, unless $q \approx q_c$,

$$\sigma^{(i)}_{\theta,q} < 0, \quad i = 1,2. \tag{37b}$$

The equivalent calculations for the discrete case depend upon the eigenvalues of the structural matrix. A specific example is carried out in Appendix 3. Stability is determined by the eigenvalues of the matrix,

$$\underset{\sim}{M}^{*}_{\theta} = I_n \otimes \underset{\sim}{W} + \tilde{\mu}\bar{\Delta} \otimes \begin{pmatrix} 1 & 0 \\ 0 & \theta^2 \end{pmatrix},$$

that is, by the eigenvalues of the matrices

$$\begin{pmatrix} \bar{k}p^{-2} + \tilde{\mu}\lambda & -ap^{-2} \\ p^{-2} & -p^{-2} + \tilde{\mu}\theta^2\lambda \end{pmatrix} \tag{38}$$

where λ is an eigenvalue of the structural matrix. Since $\lambda \leq 0$, set $q^2 = -\tilde{\mu}\lambda$. The boundary of the stability region is then given by (33a), as shown in Figure 3. The difference is, however, that not all possible values of q are attainable in the discrete case. In particular, q_c is attainable only for special values of $\tilde{\mu}$.

NONLINEAR QUESTIONS, CONTINUOUS CASE

Given an unstable situation ($\theta > \theta_c$), we now seek to provide an approximate solution for $\theta \simeq \theta_c$ that is uniformly valid in time. This is a problem in nonlinear stability theory. The theory has been under increasingly intense development for over a decade; its range of applicability has been broadened and its formal structure more clearly revealed. For recent references, see Matkowsky[33] and Kogelman and Keller.[34]

The basic idea of the approach to be used here follows that proposed in Segel[35] and Scanlon and Segel.[36] We shall employ unpublished refinements of this approach; the essentials of the resulting method were independently arrived at by Chen and Chang.[37]

We wish to provide an approximate solution to (26) subject to an initial condition of the type

$$\text{at } \bar{t} = 0, \quad \underset{\sim}{V} = \varepsilon f(\underset{\sim}{x}). \tag{39}$$

(Boundary conditions may have to be satisfied, but these need not be specified here.) In (39), ε is a small parameter that will be seen to be related to $\theta-\theta_c$. To find solutions which are $O(\varepsilon)$ uniformly in time, we set up the successive approximations procedure

$$-\partial \underset{\sim}{V}^{(1)}/\partial\bar{t} + \underset{\sim}{M}_{\theta}\underset{\sim}{V}^{(1)} = O(\varepsilon^2) \tag{40}$$

$$-\partial V^{(n)}/\partial\bar{t} + M_{\theta}V^{(n)} = N(\underset{\sim}{V}^{(n-1)}) + O(\varepsilon^{n+1}). \tag{41}$$

Since solutions are "small," it is natural to expect linear theory to provide a first approximation. Successive approximation methods notoriously generate inaccurate high order terms; the $O(\varepsilon^2)$ term in (40) and the $O(\varepsilon^{n+1})$ term in (41) indicate that the calculations will be performed only when their results are meaningful.

If (40)-(41) were adopted without modification, the leading approximation would typically be proportional to exp $(\sigma\bar{t})$ for some positive constant σ, the next approximation would have terms like exp $(2\sigma\bar{t})$, etc. The "higher" approximations would be valid over the same limited period as the leading approximation. To remedy this, we adopt the now common strategy of looking for solutions to (40)-(41) which are functions of the time variables $t_0 = \bar{t}$, $t_1 = \varepsilon\bar{t}$, $t_2 = \varepsilon^2\bar{t}$, ... We choose these functions so that the approximations $\underset{\sim}{V}^{(n)}$ are $O(\varepsilon)$ uniformly in time.

We confine our discussion to the nonlinear development of the "most dangerous sinusoidal mode," of the critical wave number. This satisfies the initial condition

$$\text{at } \bar{t} = 0, \quad \underset{\sim}{V} = \varepsilon \underset{\sim}{C}_1 \cos q_c x \tag{42}$$

where $\underset{\sim}{C}_1$ is given in (37). As shown in Appendix 2, as long as only one "dangerous mode" is present initially, essentially the same calculations serve for otherwise general $O(\varepsilon)$ initial conditions. Treatment of small but otherwise arbitrary initial conditions can in principle be accomplished by the successive approximations method, but the detailed calculations become complicated. In the two-dimensional case, the essence of the matter would be equivalent to the results of Di Prima, Eckhaus, and Segel.[38]

As the first step in the successive approximations solution, we must solve

$$-(\partial \underset{\sim}{V}^{(1)}/\partial\bar{t}) + \underset{\sim}{M}_\theta \underset{\sim}{V}^{(1)} = O(\varepsilon^2) \tag{43}$$

subject to (42). Assuming that time variation is of the multiple-scale type, we see that the solution must be of the form

$$\underset{\sim}{V}^{(1)} = \varepsilon A \underset{\sim}{C}_1 \cos q_c x \tag{44}$$

where

$$A = A(t_0, t_1, t_2, \ . \ . \ . \ ), \quad t_i = \varepsilon^i \bar{t}. \tag{45}$$

We shall require that $A = O(1)$ <u>uniformly in time</u>.

Upon substitution of (44) into (43) we obtain the requirement

$$[-\varepsilon(\partial A/\partial t_0) + \varepsilon\, \sigma^{(1)}_{\theta,q_c} A]\, \underset{\sim}{C}_1 \cos q_c x = O(\varepsilon^2).$$

To avoid an exponential solution that grows without bound when $\theta > \theta_c$ (and hence $\sigma^{(1)}_{\theta,q_c} > 0$), we take

$$\partial A/\partial t_0 = 0, \quad \sigma^{(1)}_{\theta,q_c} = O(\varepsilon). \tag{46a,b}$$

The next step is to solve

$$-(\partial \underset{\sim}{V}^{(2)}/\partial\bar{t}) + \underset{\sim}{M}_\theta \cdot \underset{\sim}{V}^{(2)} = N(\underset{\sim}{V}^{(1)}) + O(\varepsilon^3) \tag{47}$$

where from (26), (44), and (36),

$$N(\underset{\sim}{V}^{(1)}) = \tfrac{1}{2}(\hat{v}_2,\ \hat{e}_2)^T\ \varepsilon^2 A^2 (1 + \cos 2q_c x)$$
$$\hat{v}_2 = m\theta_c(a - \bar{k}m\theta_c),\quad \hat{e}_2 = 1 - m\theta_c. \tag{48}$$

The solution has the form

$$\underset{\sim}{V}^{(2)} = \varepsilon A \underset{\sim}{C}_1 \cos q_c x + \varepsilon^2 A^2 (\underset{\sim}{C}_0^{(2)} + \underset{\sim}{C}_2^{(2)} \cos 2q_c x) + O(\varepsilon^2 A^2 e^{-\sigma \bar{t}}). \tag{49}$$

The last term in the above equation stands for contributions which decay rapidly with time. Their presence is required by the initial conditions but their exact form is not crucial. For those interested, further discussion can be found in the Appendix.

In order that the middle terms in (49) provide particular solutions to (47), we must take

$$\underset{\sim}{C}_0^{(2)} = (0,\ \tfrac{1}{2}\theta_c p^3)^T = (C_0^{(2)},\ K_0^{(2)})^T, \tag{50a}$$

$$\underset{\sim}{C}_2^{(2)} = [p^3/18(m-p)]\ (4\theta_c{}^2 q_c{}^2 m^2,\ 1+4q_c{}^2)^T = (C_2^{(2)},\ K_2^{(2)})^T. \tag{50b}$$

With this, substitution of (49) into (47) yields

$$[-\varepsilon^2(\partial A/\partial t_1) + \varepsilon\sigma^{(1)}_{\theta,q_c} A]\ \underset{\sim}{C}_1 \cos q_c x = O(\varepsilon^3). \tag{51}$$

As in (46) certain conditions follow from the requirement that solutions do not grow without bound as $\bar{t} \to \infty$. In the present case, these are

$$\partial A/\partial t_1 = 0,\quad \sigma^{(1)}_{\theta,q_c} = O(\varepsilon^2). \tag{52a,b}$$

The crucial part of the calculation is at hand. We must solve

$$-(\partial \underset{\sim}{V}^{(3)}/\partial \bar{t}) + \underset{\sim}{M}_{\theta} \underset{\sim}{V}^{(3)} = \underset{\sim}{N}(\underset{\sim}{V}^{(2)}) + O(\varepsilon^4) \tag{53}$$

where from (49), the definition $\underset{\sim}{C}_1 = (m\theta_c,\ 1)^T$, and (50),

$$\underset{\sim}{N}(V^{(2)}) = \varepsilon^3 A^3\ (\hat{v}_3,\ \hat{e}_3)^T \cos q_c x + \ldots$$
$$\hat{v}_3 = a(m\theta_c K_0^{(2)} + \tfrac{1}{2}m\theta_c K_2^{(2)} + C_0^{(2)} + \tfrac{1}{2}C_2^{(2)}) - \bar{k}(2K_0^{(2)} + K_2^{(2)}) \tag{54}$$
$$\hat{e}_3 = (2m\theta_c C_0^{(2)} + m\theta_c C_2^{(2)}) - (m\theta_c K_0^{(2)} + \tfrac{1}{2}m\theta_c K_2^{(2)} + C_0^{(2)} + \tfrac{1}{2}C_2^{(2)}).$$

The dots indicate that only terms proportional to cos $q_c x$ have been retained, for only these terms will be seen to contribute to the essential equation (62) below. (Exponentially decaying contributions have also been omitted.) To pick out the "resonant" forcing function, we expand in terms of the eigenvectors. Thus

$$(\hat{v}_3,\ \hat{e}_3)^T = \gamma C_1 + \delta K_1 \tag{55}$$

where

$$\gamma = (-m^{-1}\theta_c \hat{v}_3 + \hat{e}_3)/(1-\theta_c^2). \tag{56}$$

The γ contribution is not central, for the corresponding particular solution to (53) would be proportional to the rapidly decaying quantity $\exp(\sigma^{(2)}_{\theta,q_c} t)$. The key point is that substitution into (53) of an assumed solution

$$\underset{\sim}{V}^{(3)} = \varepsilon A \underset{\sim}{C}_1 \cos q_c x + \varepsilon^2 A^2 (\underset{\sim}{C}_0^{(2)} + \underset{\sim}{C}_2^{(2)} \cos 2q_c x) + O(\varepsilon^3) \tag{57}$$

gives

$$[-\varepsilon^3(\partial A/\partial t_2) + \varepsilon\sigma^{(1)}_{\theta,q_c} A]\ \underset{\sim}{C}_1 \cos q_c x = \varepsilon^3 A^3 \gamma \underset{\sim}{C}_1 \cos q_c x + O(\varepsilon^4). \tag{58}$$

Note that the second term on the left is $O(\varepsilon^3)$ by (52b). By (52b) and (36) respectively,

$$\sigma^{(1)}_{\theta,q_c} = O(\varepsilon^2), \quad \sigma^{(1)}_{\theta,q_c} = O(\theta-\theta_c). \tag{59}$$

Thus our analysis is restricted to perturbation magnitudes ε which are related to the difference between θ and its critical value by

$$\theta-\theta_c = O(\varepsilon^2). \tag{60}$$

Referring to (36) let us write

$$\sigma^{(1)}_{\theta,q_c} = \bar{\sigma}\varepsilon^2 + O(\varepsilon^4). \tag{61}$$

By (60), (36), and (35),

$$\bar{\sigma} = O(1), \quad \text{sgn}\ \bar{\sigma} = \text{sgn}\ (\theta-\theta_c).$$

Using (61), from (58) we have to lowest order

$$\partial A/\partial t_2 = \bar{\sigma}A - \gamma A^3. \tag{62}$$

In the present instance, after some lengthy algebraic manipulations which are characteristic of nonlinear stability theory, one arrives at the expression

$$\gamma = p^4(30p^2-22mp-23m^2)/[36m(m-p)\{(m-p)^2-1\}].$$

Given the restrictions $m > p$ and $m^2-p^2 < 1$ specified earlier, the "Landau constant" γ must be positive.

From the successive approximations, we therefore draw the following conclusions concerning the amplitude εA of the Fourier mode of critical wavelength $2\pi/q_c$, when $\theta > \theta_c$.

(i) Equation (46a) shows that A does not change on the fast O(1) time scale.

(ii) Equation (52) shows that A also does not change on the

intermediate $O(\varepsilon^{-1})$ time scale.

(iii) According to linear theory A grows exponentially like $\exp(\bar{\sigma}\varepsilon^2)t$, but (62) shows that on an $O(\varepsilon^{-2})$ time scale A actually levels off to an O(1) equilibrium value $(\bar{\sigma}/\gamma)^{\frac{1}{2}}$. Thus the amplitude of the critical mode is $O(\varepsilon)$ uniformly in time. The situation is illustrated in Figure 4(a).

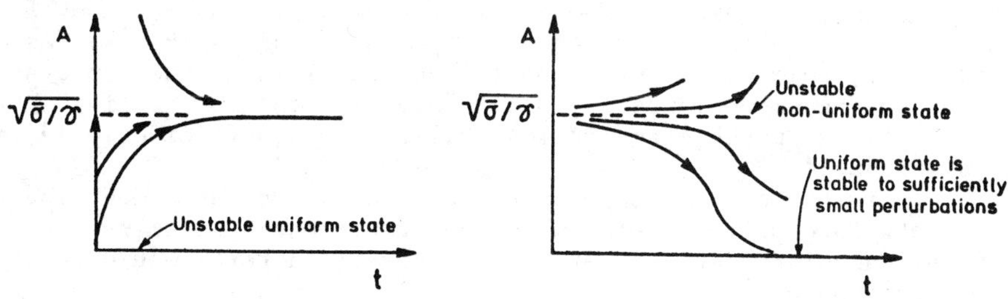

Figure 4(a). Possible curves A(t) when $\gamma > 0$, for $\theta > \theta_c$ (so $\bar{\sigma} > 0$). $A = \sqrt{\bar{\sigma}/\gamma}$ is a stable equilibrium state.

Figure 4(b). Possible curves A(t) when $\gamma < 0$, for $\theta < \theta_c$ (so $\bar{\sigma} < 0$). $A = \sqrt{\bar{\sigma}/\gamma}$ is an unstable equilibrium state.

NONLINEAR CALCULATIONS, DISCRETE CASE

The calculations in the discrete case are similar to those just given, and are summarized in Appendix 3. When $\tilde{\mu}$ is such that the corresponding θ_c is not too large, then $\gamma > 0$ and results are analogous to the continuous case. But γ is certainly negative when $\tilde{\mu} \ll 1$ or when $\tilde{\mu} \gg 1$ and $\bar{k} > a/2$.

Now a new non-uniform equilibrium with $A^2 = \bar{\sigma}/\gamma$ exists only when $\theta < \theta_c$, a situation where the basic uniform state is stable according to linear theory. From (62) we see that solutions behave as in Figure 4(b). Thus, when applied to the uniform state, appropriate perturbations whose magnitudes are slightly larger than $\sqrt{\bar{\sigma}/\gamma}$ will grow even though linear theory predicts stability.

CONCLUSIONS

In a homogeneous continuous environment, random dispersal can destabilize a predator-prey interaction in the presence of an Allee (autocatalytic) effect in the prey population, provided that the predator dispersal ability is sufficiently high with respect to that of the prey. As perturbations grow, nonlinear effects come into play and bring about a new steady but now spatially inhomogeneous species distribution.

Similar effects appear in the discrete spatial case (patchy environment): random dispersal can again destabilize in the presence of an Allee effect, and a spatially inhomogeneous pattern can ensue. As the number of patches becomes large, the instability conditions approach those of the continuous case. But in a fixed patchy environment, in contrast to the continuous case, whatever the ratio of dispersal rates the homogeneous situation will be stable to infinitesimal perturbations if prey dispersal is sufficiently vigorous.

In the continuous case, only values of θ near the value θ_c of Figure 3 come into the purview of our analysis. (If $\theta >> \theta_c$, disturbances to the basic state grow to such a large amplitude that our theory is invalid.) The discrete case permits selection of arbitrary values of $\tilde{\mu}$ and hence of the corresponding $\theta_c^2(\tilde{\mu})$. As we have seen, this extra degree of freedom can lead to the appearance of a new phenomenon, the so-called "sub-critical instability" where a uniform two-patch state is destabilized by a sufficiently large disturbance.

It is beyond the scope of our present analysis to describe the ultimate fate of the system where no stable equilibrium is yielded by the calculations. Higher order interaction terms would be important here.

Returning to our discussion of the continuous case, we observe that the combination of prey autocatalysis and high predator dispersal ability introduces a short-range activation--long-range inhibition coupling into the system which results in destabilization of uniform spatial patterns.[39,1] Stable patterns appear, in which the prey, especially, tends to be clumped [because $m\theta_c > 1$ in the vector $\underset{\sim}{C}_1$ of (37a)]. Individual prey benefit from clumping due to reduced predator efficiency at culling them. Overall predator density will be higher with the clumped pattern than it would be under the uniform distribution of prey and predator, ignoring dispersal [see (50a)].

The reverse situation is also of interest. With highly mobile prey and an Allee effect in the predator population (due perhaps to enhanced mating or group hunting), the resultant pattern will exhibit high predator clumping (plus, of course, some non-uniformity in prey dispersion).

The principal conclusion of this exercise is that <u>patchiness in species distributions may arise even in a homogeneous environment through species interactions</u>. Okubo[40] has suggested that exactly the diffusion-induced instability we discuss here is one possible mechanism underlying the patchy distribution of oceanic phytoplankton. Steele[41] has also discussed the development of patchiness when the homogeneous steady-state is unstable even with diffusion.

In general, the linear perturbation equations for the homogeneous case have the form

$$\partial v/\partial t = a_{11}v + a_{12}e, \quad \partial e/\partial t = a_{21}v + a_{22}e.$$

The self-reinforcement coefficient a_{11} of the victims is positive while that of the exploiters (a_{22}) is negative [compare (26)]. Segel and Jackson[2] showed that the interaction can result in instability of the uniform steady state if relatively rapid diffusion causes too rapid decay of exploiter perturbations (Levin[3] has analyzed a complementary situation). When this happens, in spite of coupling between exploiter and victim perturbations, the exploiter perturbation dies out before it can exert enough of a stabilizing effect. Thus instability sets in when θ^2, the ratio of exploiter diffusivity ν to victim diffusivity μ, exceeds a critical value θ_c^2.

The present analysis shows that when $\theta > \theta_c$ a two-dimensional unstable perturbation does not continue to intensify, but rather levels off to a new, spatially inhomogeneous, steady state. Generally, there is a distortion of the periodic spatial distribution due to the generation of the cos $2q_cy$ harmonic, and a change in the average concentrations. It is these changes that alter the overall predator-prey interaction to limit the perturbation growth and bring about a new equilibrium. In the present instance, we see from (50a) that the average victim concentration is unaltered by the nonlinear effects, but the average exploiter concentration is elevated--so at least one of the nonlinear effects provides the relatively enhanced exploiter effectiveness which can be anticipated from the existence of a new steady state.

Our analysis has been restricted to perturbations which are a function of a single spatial coordinate. When two-dimensional effects are treated we expect from the formalism in problems with similar structure (see for example the paper of Scanlon and Segel[36]) that two new phenomena may arise. Firstly, a sufficiently large perturbation to the uniform state may grow if θ is close enough to θ_c, even though $\theta < \theta_c$, a result which also appears in the discrete case (Appendix 3). Secondly, the ultimate steady state may be able to fill the plane by repetition of a pattern which can be regarded as having a hexagonal boundary, although (depending on the sign of a coefficient yielded by the analysis) it may appear either as a repetition of roughly circular density concentrations or of a netlike pattern of relatively high density.

In a simple model of slime mold aggregation, Nanjundiah's[42] calculations show that once the linear stability limits have been exceeded, aggregation continues until all organisms are concentrated at discrete points. Somewhat more general models can display a steady non-homogeneous structure of the type found here.[43] In slime mold, instability is brought about by chemotaxis in spite of random dispersal. Here, dispersal "causes" instability. That the latter type of instability does not lead to intense concentrations of organisms might be anticipated from the fact that at such concentrations the subtle decoupling effects of random dispersal are masked by its more straightforward role as a disperser of local accumulations.

It should be kept in mind that the dispersal studied here is random. Probably a more prevalent type of dispersal is tactic movement toward some substances (e.g. food) and away from others. (For example, the investigations of Bull and Mitchell,[44] Adler,[45,46] and others, show the importance of tactic motions in bacterial ecology.)

Taxis can be modelled analogously to the treatment of chemotaxis by Keller and Segel's[47] initial study of slime mold aggregation. The development of spatial inhomogeneity via instability is to be expected in tactic organisms, for evolution is expected to lead to increased tactic sensitivity, and too much sensitivity causes aggregation. One is tempted to conjecture that instability due to dispersal, random and tactic, is a significant factor in producing the spatial inhomogeneity that enhances coexistence of species and possibly speciation.

APPENDIX 1

In this section we detail the region of instability in the discrete case (we do not non-dimensionalize). We assume $\alpha < 0 < \delta$. Utilizing (13), we see that this region I is given by

$$I = \bigcup_{\lambda_i \varepsilon \Omega} [(\tilde{\mu},\tilde{\nu}):(\lambda_i\tilde{\mu} - \alpha\hat{V})(\lambda_i\tilde{\nu} - \delta\hat{E}) + \beta\gamma\hat{E}\hat{V} < 0]$$

where $\Omega = [\lambda_i]$ is the set of eigenvalues of Δ. Since Δ is real and symmetric, the eigenvalues λ_i are real and may be arranged in descending order.

$$0 = \lambda_1 \geq \lambda_2 \ . \ . \ . \geq \lambda_n .$$

For any particular choice of $\lambda_i \neq 0$, the set

$$(\lambda_i\tilde{\mu} - \alpha\hat{V})(\lambda_i\tilde{\nu} - \delta\hat{E}) + \beta\gamma\hat{E}\hat{V} = 0$$

describes a hyperbola which cuts the positive $\tilde{\nu}$ axis at

$$\tilde{\nu}_i = [(\alpha\delta + \beta\gamma)/\alpha]E\cdot(1/\lambda_i) = A/\lambda_i,$$

in which $A < 0$ and $\lambda_i \leq 0$; and $\tilde{\nu} \to +\infty$ as $\tilde{\mu} \to \alpha\hat{V}/\lambda_i = B/\lambda_i$ from below. Recall that $B < 0$. As i increases (beyond the point where the first non-zero eigenvalue is attained), λ_i becomes more negative, $\tilde{\nu}_i$ smaller, and B/λ_i smaller. This situation is shown in Figure 1, in which the region of instability is shaded. It is a straightforward exercise to show that each of the hyperbolas is tangent to the line

$$\tilde{\nu} = (\hat{E}/\hat{V})\tilde{\mu}(\sqrt{\alpha\delta + \beta\gamma} + \sqrt{\beta\gamma})^2/\alpha^2,$$

which corresponds to the boundary of the region of instability in the continuous case.

As a discrete system becomes more and more reticulated, approaching a continuum, the eigenvalues of Δ become closer to one another and the region in Figure 1 tends to that corresponding to the continuous case.

APPENDIX 2
SOME DETAILS CONCERNING RAPIDLY DECAYING PARTS OF THE SOLUTION

Compared to $\underset{\sim}{V}^{(1)}$, the second approximation $\underset{\sim}{V}^{(2)}$ is more accurate largely because of additional $O(\varepsilon^2A^2)$ terms which alter the spatial average and which add a second harmonic to the perturbation. But if these terms alone appeared in (49), the initial condition

(39) would not be satisfied. To deal with this for the modification of the average, we must add to the solution

$$-\varepsilon^2 A^2 [\beta_1 \underset{\sim}{C}^{(1)}_{\theta,0} \exp (\sigma^{(1)}_{\theta,0} t) + \beta_2 \underset{\sim}{C}^{(1)}_{\theta,0} \exp (\sigma^{(2)}_{\theta,0} t)]$$

where β_1 and β_2 are constants determined by the requirement

$$\underset{\sim}{C}^{(2)}_0 = \beta_1 \underset{\sim}{C}^{(1)}_{\theta,0} + \beta_2 \underset{\sim}{C}^{(2)}_{\theta,0}.$$

Here $\underset{\sim}{C}^{(1)}_{\theta,q}$ and $\underset{\sim}{C}^{(2)}_{\theta,q}$ are eigenvectors defined by

$$\underset{\sim}{M}_{\theta,q} \underset{\sim}{C}^{(i)}_{\theta,q} = \sigma^{(i)}_{\theta,q} \underset{\sim}{C}^{(i)}_{\theta,q}; \quad i = 1,2.$$

The second harmonic is treated in the same fashion.

The presence of the term proportional to

$$\varepsilon^2 A^2 \exp (\sigma^{(1)}_{\theta,0} t)$$

in $\underset{\sim}{V}^{(2)}$ leads, among others, to a term proportional to

$$\varepsilon^3 A^3 \exp (\sigma^{(1)}_{\theta,0} t) \cos q_c x$$

in $\underset{\sim}{N}(\underset{\sim}{V}^{(2)})$. The principal resulting contribution to the solution is proportional to $\varepsilon^3 A^3 \exp (\sigma^{(1)}_{\theta,0} t)$. The exponential factor becomes entirely negligible in an $O(1)$ time interval; the interesting change in A takes place over a much longer $O(\varepsilon^{-2})$ interval and is barely affected. (Bear in mind that $\sigma^{(i)}_{\theta,q}$ is $O(1)$ and negative unless $q \approx q_c$.) Again, other terms are treated in the same fashion. Indeed, similar considerations show that more general initial conditions than (39), including the possible presence of a term proportional to $\underset{\sim}{K}_1 \cos q_c x$, would also add to the solution only relatively uninteresting, rapidly decaying, terms.

APPENDIX 3
OUTLINE OF NONLINEAR STABILITY CALCULATIONS IN THE DISCRETE CASE

Strictly for illustrative purposes, we carry through in outline form the nonlinear calculations for a discrete model of special form ($n = 2$) to show that the general phenomenon of the development of spatial pattern in a homogeneous environment holds true. Increasing n would not change the qualitative conclusions.

The situation we describe thus involves two patches, and without loss of generality we assume that the connectivity $c_{12} = \bar{k} = \alpha/\gamma$. Any variation from this can be absorbed into the parameters $\hat{\mu}$, $\hat{\nu}$. Thus

$$\underset{\sim}{\bar{\Delta}} = \begin{pmatrix} -1 & 1 \\ 1 & -1 \end{pmatrix}.$$

Moreover, recall that

(39) would not be satisfied. To deal with this for the modification of the average, we must add to the solution

$$-\varepsilon^2 A^2 [\beta_1 \underset{\sim}{C}^{(1)}_{\theta,0} \exp (\sigma^{(1)}_{\theta,0} t) + \beta_2 \underset{\sim}{C}^{(1)}_{\theta,0} \exp (\sigma^{(2)}_{\theta,0} t)]$$

where β_1 and β_2 are constants determined by the requirement

$$\underset{\sim}{C}^{(2)}_0 = \beta_1 \underset{\sim}{C}^{(1)}_{\theta,0} + \beta_2 \underset{\sim}{C}^{(2)}_{\theta,0}.$$

Here $\underset{\sim}{C}^{(1)}_{\theta,q}$ and $\underset{\sim}{C}^{(2)}_{\theta,q}$ are eigenvectors defined by

$$\underset{\sim}{M}_{\theta,q} \underset{\sim}{C}^{(i)}_{\theta,q} = \sigma^{(i)}_{\theta,q} \underset{\sim}{C}^{(i)}_{\theta,q}; \quad i = 1,2.$$

The second harmonic is treated in the same fashion.

The presence of the term proportional to

$$\varepsilon^2 A^2 \exp (\sigma^{(1)}_{\theta,0} t)$$

in $\underset{\sim}{V}^{(2)}$ leads, among others, to a term proportional to

$$\varepsilon^3 A^3 \exp (\sigma^{(1)}_{\theta,0} t) \cos q_c x$$

in $\underset{\sim}{N}(\underset{\sim}{V}^{(2)})$. The principal resulting contribution to the solution is proportional to $\varepsilon^3 A^3 \exp (\sigma^{(1)}_{\theta,0} t)$. The exponential factor becomes entirely negligible in an O(1) time interval; the interesting change in A takes place over a much longer $O(\varepsilon^{-2})$ interval and is barely affected. (Bear in mind that $\sigma^{(1)}_{\theta,q}$ is O(1) and negative unless $q \approx q_c$.) Again, other terms are treated in the same fashion. Indeed, similar considerations show that more general initial conditions than (39), including the possible presence of a term proportional to $\underset{\sim}{K}_1 \cos q_c x$, would also add to the solution only relatively uninteresting, rapidly decaying terms.

APPENDIX 3
OUTLINE OF NONLINEAR STABILITY CALCULATIONS IN THE DISCRETE CASE

Strictly for illustrative purposes, we carry through in outline form the nonlinear calculations for a discrete model of special form (n = 2) to show that the general phenomenon of the development of spatial pattern in a homogeneous environment holds true. Increasing n would not change the qualitative conclusions.

The situation we describe thus involves two patches, and without loss of generality we assume that the connectivity $c_{12} = \bar{k} = \alpha/\gamma$. Any variation from this can be absorbed into the parameters $\tilde{\mu}$, $\tilde{\nu}$. Thus

$$\bar{\underset{\sim}{\Delta}} = \begin{pmatrix} -1 & 1 \\ 1 & -1 \end{pmatrix}.$$

Moreover, recall that

$$\underset{\sim}{W} = p^{-2}\begin{pmatrix}\bar{k} & -a \\ 1 & -1\end{pmatrix} \text{ and } \underset{\sim}{V} = (v_1, e_1, v_2, e_2)^T$$

and define

$$\underset{\sim}{P} = \begin{pmatrix}1 & 0 \\ 0 & \theta^2\end{pmatrix}.$$

Then (29)

$$\underset{\sim}{M}_\theta^* = \underset{\sim}{I}_2 \otimes \underset{\sim}{W} + \tilde{\mu}\underset{\sim}{\bar{\Delta}} \otimes \underset{\sim}{P};$$

and as before

$$\underset{\sim}{N}(\underset{\sim}{V}) = (ae_1v_1 - \bar{k}v_1^2, e_1^2 - e_1v_1, ae_2v_2 - \bar{k}v_2^2, e_2^2 - e_2v_2)^T.$$

The basic problem involves solutions to

$$\underset{\sim}{L}\,\underset{\sim}{V} = -(d\underset{\sim}{V}/dt) + \underset{\sim}{M}_\theta^*\underset{\sim}{V} = \underset{\sim}{N}(\underset{\sim}{V}).$$

The eigenvalues of $\bar{\Delta}$ are 0 and -2, and it is easily shown that the region of instability corresponds to that given for the continuous case in Figure 3, but with q replaced by $\sqrt{2\tilde{\mu}}$ (Figure 5).

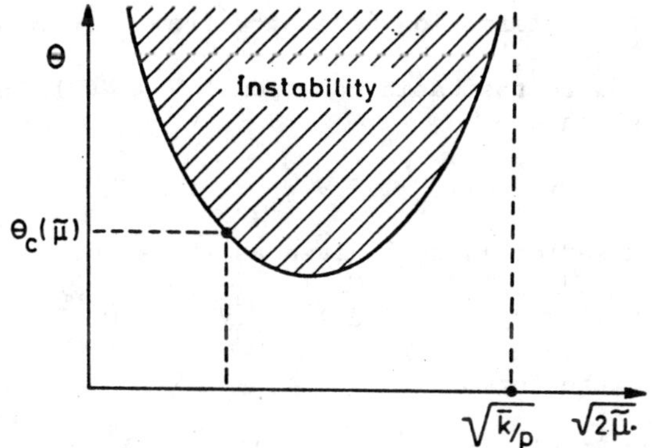

Figure 5. Stability diagram in the discrete case.

We assume $\tilde{\mu}$ to be given and define $\theta_c^2(\tilde{\mu})$ to be the diffusivity ratio where destabilization occurs <u>for that</u> $\tilde{\mu}$. Note that there is no reason to confine our attention to the region around the bottom of the neutral curve, as in the continuous case, and in the general case we find that

$$\theta_c^{\,2}(\tilde{\mu}) = (1 + 2\tilde{\mu})/(2\tilde{\mu}r_2),$$

where $r_2 = \bar{k} - 2\tilde{\mu}p^2$.

The linear problem is treated first, and the appropriate linearization matrix is $\hat{M}_\theta$. Let σ_3, σ_4 be the roots of

$$\sigma^2 - [(\bar{k}-1)/p^2]\sigma + [(a-\bar{k})/p^4] = 0,$$

i.e. the eigenvalues of $\underset{\sim}{W}$, with eigenvectors $\underset{\sim}{W}_3$, $\underset{\sim}{W}_4$. Then σ_3, σ_4 are eigenvalues of $\underset{\sim}{M}_\theta^*$, with eigenvectors $(1, 1)^T \otimes \underset{\sim}{W}_3$ and $(1, 1)^T \otimes \underset{\sim}{W}_4$.

Consider next the matrix

$$\underset{\sim}{T} = p^{-2}\begin{pmatrix} \bar{k} & -a \\ 1 & -1 \end{pmatrix} - 2\tilde{\mu}\begin{pmatrix} 1 & 0 \\ 0 & \theta^2 \end{pmatrix}.$$

Near det T = 0, its eigenvalues are

$$\sigma_1 = m_1(\theta-\theta_c) + O(\theta-\theta_c)^2, \quad \sigma_2 = m_2 + O(\theta-\theta_c),$$

where $m_1 > 0$ and $m_2 < 0$ are certain O(1) constants. The eigenvectors are

$$\underset{\sim}{W}_1 = (r_1, 1)^T + O(\theta-\theta_c), \quad \underset{\sim}{W}_2 = (r_2, 1)^T + O(\theta-\theta_c),$$

where $r_1 = a/r_2$. σ_1 and σ_2 are eigenvalues of $\underset{\sim}{M}_\theta^*$ with eigenvectors

$$(1, -1)^T \otimes \underset{\sim}{W}_1 \text{ and } (1, -1)^T \otimes \underset{\sim}{W}_2.$$

This completes the information necessary from the linear approximation.

We return now to the basic equation $\underset{\sim}{L}\,\underset{\sim}{V} = \underset{\sim}{N}(\underset{\sim}{V})$, and introduce a perturbation given by

$$\underset{\sim}{V} = \varepsilon(1, -1)^T \otimes \underset{\sim}{W}_1 \text{ at } t = 0.$$

The first approximation is then given as the solution to

$$\underset{\sim}{L}\,\underset{\sim}{V}^{(1)} = -(d\underset{\sim}{V}^{(1)}/dt) + \underset{\sim}{M}_\theta^*\underset{\sim}{V}^{(1)} = O(\varepsilon^2).$$

Its solution has the form

$$\underset{\sim}{V}^{(1)} = \varepsilon A(t_0, t_1, \ldots)\ (1, -1)^T \otimes \underset{\sim}{W}_1.$$

Note that

$$\underset{\sim}{L}\,\underset{\sim}{V}^{(1)} = -\varepsilon(\partial A/\partial t_0)\ (1, -1)^T \otimes \underset{\sim}{W}_1 + \varepsilon A\underset{\sim}{M}_\theta^*[(1, -1)^T \otimes \underset{\sim}{W}_1] + \ldots$$

But, since we are at $\theta = \theta_c$,

$$\underset{\sim}{M}_\theta^*[(1, -1)^T \otimes \underset{\sim}{W}_1] = \sigma_1(1, -1)^T \otimes \underset{\sim}{W}_1.$$

Thus

$$[-\varepsilon(\partial A/\partial t_0) + \varepsilon\sigma_1 A] = O(\varepsilon^2),$$

and so

$$\partial A/\partial t_0 = 0, \quad \sigma_1 = O(\varepsilon).$$

The second approximation satisfies

$$-(d\underset{\sim}{V}^{(2)}/dt) + \underset{\sim}{M}^{*}_{\theta}\underset{\sim}{V}^{(2)} = \underset{\sim}{N}(\underset{\sim}{V}^{(1)}) + O(\varepsilon^3).$$

Here,

$$N(\underset{\sim}{V}^{(1)}) = (1, 1)^T \otimes (ar_1 - \bar{k}r_1{}^2, 1 - r_1)^T \varepsilon^2 A^2.$$

The solution is

$$\underset{\sim}{V}^{(2)} = \varepsilon A(1, -1)^T \otimes \underset{\sim}{W}_1 + \varepsilon^2 A^2 (1, 1)^T \otimes (r_3, s_3)^T + \ldots,$$

where

$$(r_3, s_3)^T = \underset{\sim}{W}^{-1}(ar_1 - \bar{k}r_1{}^2, 1 - r_1).$$

Note that

$$\underset{\sim}{L}\,\underset{\sim}{V}^{(2)} = [-\varepsilon^2(\partial A/\partial t_1) + \varepsilon\sigma_1 A](1, -1)^T \otimes \underset{\sim}{W}_1 + \underset{\sim}{N}(\underset{\sim}{V}^{(1)}) + O(\varepsilon^3);$$

and so

$$\partial A/\partial t_1 = 0, \quad \sigma_1 = O(\varepsilon^2).$$

Finally,

$$-(d\underset{\sim}{V}^{(3)}/dt) + \underset{\sim}{M}^{*}_{\theta}\underset{\sim}{V}^{(3)} = \underset{\sim}{N}(\underset{\sim}{V}^{(2)}) + O(\varepsilon^4).$$

Note that

$$\underset{\sim}{N}(\underset{\sim}{V}^{(2)}) = (1, 1)^T \otimes (ar_1 - \bar{k}r_1{}^2, 1 - r_1)^T \varepsilon^2 A^2 +$$
$$2\varepsilon^3 A^3 (1, -1)^T \otimes [a(r_3+s_3r_1) - 2\bar{k}r_1r_3, 2s_3 - (r_3+s_3r_1)]^T + O(\varepsilon^4).$$

At this stage it is not necessary to retain terms proportional to $(1, 1)^T \otimes$ anything, since such terms will not contribute to the crucial part of the solution proportional to the "most dangerous" eigenvector $(1, -1)^T \otimes \underset{\sim}{W}_1$. Thus we can write

$$\underset{\sim}{N}(\underset{\sim}{V}^{(2)}) = \varepsilon^3 A^3 (1, -1)^T \otimes (r_4, s_4)^T + \ldots,$$

where

$$r_4 = 2[a(r_3 + s_3r_1) - 2\bar{k}r_1r_3], \quad s_4 = 2[2s_3 - (r_3 + s_3r_1)].$$

Moreover, to obtain the part of the solution proportional to $(1, -1)^T \otimes \underset{\sim}{W}_1 \equiv (1, -1)^T \otimes (r_1, 1)^T$, it suffices to consider the term $\gamma(r_1, 1)^T$ in

$$(r_4, s_4)^T = \gamma(r_1, 1)^T + \delta(r_2, 1)^T,$$

where

$$\gamma = (r_4 - r_2s_4)/(r_1 - r_2).$$

Thus we seek to solve

$$\underset{\sim}{L}\,\underset{\sim}{V}^{(3)} = -(d\underset{\sim}{V}^{(3)}/dt) + \underset{\sim}{M}^{*}_{\theta}\underset{\sim}{V}^{(3)} = \varepsilon^3 A^3 \gamma (1,\ -1)^T \otimes \underset{\sim}{W}_1 + \ . \ . \ . + O(\varepsilon^4).$$

Assume a solution

$$\underset{\sim}{V}^{(3)} = \varepsilon A(1,\ -1)^T \otimes \underset{\sim}{W}_1 + \varepsilon^2 A^2 (1,\ 1)^T \otimes (r_3,\ s_3)^T + \ . \ . \ . + O(\varepsilon^3).$$

Then

$$L\,\underset{\sim}{V}^{(3)} = [-\varepsilon^3(\partial A/\partial t_2) + \varepsilon\sigma_1 A](1,\ -1)^T \otimes \underset{\sim}{W}_1 \ . \ . \ . + O(\varepsilon^4)$$

$$= \varepsilon^3 A^3 \gamma (1,\ -1)^T \otimes \underset{\sim}{W}_1 + \ . \ . \ . + O(\varepsilon^4).$$

Thus

$$-\varepsilon^3(\partial A/\partial t_2) + \varepsilon\sigma_1 A = \varepsilon^3 A^3 \gamma.$$

Now

$$\sigma_1 = \bar{\sigma}\varepsilon^2 + O(\varepsilon^4),$$

and so

$$\partial A/\partial t_2 = \bar{\sigma} A - \gamma A^3.$$

We wish information on the sign of γ. Since $0 < \bar{k} < 1,a$, it follows that $r_2 < 1,a$ and $r_1 = a/r_2 > 1,a$; so that $r_1 > r_2$. Thus $\text{sgn}\ \gamma = \text{sgn}\ (r_4 - r_2 s_4)$. We define

$$z = a(1-r_1)^2/[(a-\bar{k})r_1^{\ 2}].$$

Then

$$r_1(r_4 - r_2 s_4)/2(a-\bar{k})^2 r_1^{\ 4} = -z^2 + z r_1^{\ -2}(3r_1^{\ 2} + r_1 + 2) - 2.$$

Near the minimum of the neutral curve of Figure 5, where $2\tilde{\mu} \approx (m-p)/p$, and $\theta_c \approx 1/(m-p)$, the calculations are essentially the same as in the continuous case and $\gamma > 0$. But for $0 < \tilde{\mu} \ll 1$, it is easily seen that $\gamma < 0$.

ACKNOWLEDGMENTS

L. A. Segel was supported by the Army Research Office and S. A. Levin by National Science Foundation Grant GP 33031. The research was also supported by the U.S.-Israel BiNational Science Foundation.

REFERENCES

1. A. Turing, Phil. Trans. Roy. Soc. B237, 37 (1952).
2. L. A. Segel and J. L. Jackson, J. Theoret. Biol. 37, 545 (1972).
3. S. A. Levin, Am. Nat. 108, 207 (1974).
4. R. T. Paine, Am. Nat. 100, 65 (1966).
5. R. T. Paine, Ecology 52, 1096 (1971).
6. J. W. Porter, Am. Nat. 106, 487 (1972).

7. J. L. Harper, Brookhaven Symp. 22, 48 (1969).
8. J. L. Brooks and S. I. Dodson, Science 150, 28 (1965).
9. T. M. Zaret, Ecology 53, 248 (1972).
10. T. M. Zaret and R. T. Paine, Science 182, 449 (1973).
11. D. Pimentel, Science 159, 1432 (1968).
12. A. J. Lotka, Elements of Mathematical Biology (Dover, N.Y., 1925).
13. V. Volterra, Mem. Accad. Nazionale Lincei (ser 6) 2, 31 (1926).
14. V. Volterra, Leçons sur la théorie mathématique de la lutte pour la vie (Gauthier-Villars, Paris, 1931).
15. G. F. Gause, The Struggle for Existence (Williams & Wilkins, Baltimore, 1934).
16. L. S. Luckinbill, Ecology 55, 1142 (1974).
17. C. B. Huffaker, Hilgardia 27, 343 (1958).
18. M. L. Rosenzweig, Am. Nat. 103, 81 (1969).
19. C. J. Krebs, Ecology: The Experimental Analysis of Distribution and Abundance (Harper & Row, N.Y., 1972).
20. J. Maynard Smith, Models in Ecology (Cambridge Univ. Press, Cambridge, England, 1974).
21. C. S. Holling, Mem. Entomol. Soc. Can. 45, 1 (1965).
22. W. C. Allee, The Social Life of Animals (Heinemann, London, 1939).
23. R. M. May, Science 177, 900 (1972).
24. A. N. Kolmogorov, Giorn. Inst. Ital. Attuari 7, 74 (1936).
25. A. T. Winfree, J. Theoret. Biol. 16, 15 (1967).
26. G. E. Hutchinson, Am. Nat. 93, 145 (1959).
27. J. G. Skellam, Biometrika 38, 196 (1951).
28. J. H. Van der Meer, J. Theoret. Biol. 41, 161 (1973).
29. H. G. Othmer and L. E. Scriven, J. Theoret. Biol. 32, 507 (1971).
30. H. G. Othmer and L. E. Scriven, J. Theoret. Biol. 43, 87 (1974).
31. B. Friedman, Proc. Conf. on Differential Equations (Univ. Maryland, College Park, 1956), p. 209.
32. M. Marcus and H. Minc, A Survey of Matrix Theory and Matrix Inequalities (Allyn & Bacon, Inc., Boston, 1964).
33. B. M. Matkowsky, Bull. Amer. Math. Soc. 76, 620 (1970).
34. S. Kogelman and J. B. Keller, SIAM J. Appl. Math. 20, 619 (1971).
35. L. A. Segel, in Non-equilibrium Thermodynamics, Variational Techniques and Stability, R. Donnelly, R. Herman, and I. Prigogine, Eds. (Univ. Chicago Press, 1966).
36. J. W. Scanlon and L. A. Segel, J. Fluid Mechanics 30, 149 (1967).
37. Y. M. Chen and J. S. Chang, SIAM J. Appl. Math. 23, 411 (1972).
38. R. C. Di Prima, W. Eckhaus, and L. A. Segel, J. Fluid Mechanics 49, 705 (1971).
39. A. Gierer and H. Meinhardt, Kybernetic 12, 30 (1972).
40. A. Okubo, Tech. Rep. 86, Chesapeake Bay Inst., Johns Hopkin Univ. (1974).
41. J. H. Steele, Coastal Upwelling Ecosystems Analysis Newsletter 2(4), 3 (1973).

42. V. Nanjundiah, J. Theoret. Biol. 42, 63 (1973).
43. E. Pate and L. A. Segel, unpublished.
44. W. Bell and R. Mitchell, Biol. Bull. 143, 265 (1972).
45. J. Adler, Science 153, 708 (1966).
46. J. Adler, Science 166, 1588 (1969).
47. E. F. Keller and L. A. Segel, J. Theoret. Biol. 26, 399 (1970).

Dr. Segel is Professor of Applied Mathematics at the Weizmann Institute, and Research Professor of Mathematics at Rensselaer Polytechnic Institute. He has been a Visiting Professor at the Sloan-Kettering Institute and MIT. Recently he has been mainly concerned with mathematical problems in biology and in nonlinear hydrodynamic stability theory. He met Julius at the Weizmann Institute in 1971.

Dr. Levin is Chairman of the Section of Ecology and Systematics at Cornell University. His recent work has concerned problems in mathematical ecology.

FLUCTUATIONS IN CONTINUOUS SYSTEMS

N.G. van Kampen
Institute for Theoretical Physics of the University
Utrecht, Netherlands

ABSTRACT

The density of particles in space or in phase space varies in time through continuous flow and through discrete jumps, due to collisions, chemical reactions, radio-active decay, or births and deaths. A method is developed for computing the fluctuations arising from the random jumps and influenced by the flow. It equally applies to equilibrium, stationary states, and time-dependent situations, both in linear and nonlinear systems. The method is based on the master equation and does not require the additional assumptions needed for the Langevin approach, although the correct form of the Langevin force can be deduced from it a posteriori. Factorial cumulants turn out to be a convenient tool. As applications the fluctuations inherent in the diffusion process are computed and a controversial chemical reaction is discussed.

1. INTRODUCTION

The validity of macroscopic transport laws or rate equations is subject to restrictions due to the discrete nature of matter. The deviations from the macroscopic behavior show up in fluctuations. In principle all properties of the fluctuations are implied in the microscopic equations of motion of all particles, but the road from the microscopic starting point to the observed fluctuations is long and hazardous. It involves all the difficulties connected with the appearence of irreversibility, and cannot, therefore, dispense with some statistical assumptions. In practice one starts from an intermediate level of description, more detailed than the macroscopic picture, but less detailed than the microscopic one, and introduces some suitable and physically reasonable assumptions about the behavior on that level. "Mesoscopic" is a suitable term for this approach.

One way of computing fluctuations mesoscopically consists in supplementing the macroscopic equations with a fluctuating Langevin term and to assume certain stochastic properties for it. However, the physical justification for these assumptions is dubious, unless the Langevin term corresponds to an actual force, as in the Brownian particle case. Moreover, in time dependent situations the Langevin method can only be saved by allowing the stochastic properties to depend on the special solution under consideration.[5]

More satisfactory is that mesoscopic approach that starts by setting up the master equation. It equally applies to stationary and time dependent states, and often requires no other assumptions than have already been used in establishing the macroscopic equations. All that is needed is to frame them in an equation for the probability distribution, i.e., the master equation. One is then left, however, with the problem of solving it.

Exact solutions of the master equation are available only in a restricted class of problems, to be identified in Sec.2 as "linear" master equations. For nonlinear master equations an approximation method has been developed, which produces solutions in the form of an expansion in reciprocal powers of a parameter measuring the size of the system. This method was first formulated for the case of a single fluctuating variable [1,2,3], and subsequently generalized to the multivariate case [4,5]. The present article studies the case of infinitely many variables; more precisely, the case that the fluctuating quantity is itself a function of a continuously varying coordinate. Examples of such quantities are the density of particles in space, or in one-particle phase space.

Density fluctuations in equilibrium can of course be calculated by the usual phase space methods [6], an example is given in Sec.7. Non-equilibrium fluctuations, however, in continuous systems have appeared in the literature in connection with various special problems, such as the derivation of the Onsager relations [7], the theory of noise in electron beams [8], and the study of hot electrons [11,12,13]; and in connection with transport processes [14,15] generalized hydrodynamics [16], the electromagnetic field in dielectrics [17], lasers [18], chemical reactions [19 - 20], and even sociology [23,18]. Often, however, recourse had to be taken to ad hoc methods, involving assumptions and simplifications (optimistically called "approximations"), which sometimes led to ambiguities and controversies [24,10,19 - 22]. A more general systematic, if rather cumbersome, treatment was given by Van Vliet [25], who obtained several of the results that will be derived here. His method, however, is based on the Langevin equation and is therefore subject to the objections mentioned above. Recently Mori [31] has treated continuous variables by Fourier transformation, which reduces them to a set of discrete variables.

Our aim is to provide a general method that permits to calculate without ambiguities the fluctuations in stationary or time-dependent states of linear and nonlinear systems with continuous variables, using no other assumptions than those that are indispensable for a mesoscopic treatment. Of course this does not mean that we claim to have resolved <u>all</u> difficulties; in particular the effect of space charge has yet to be worked out.

The difficulty in dealing with continuous variables is that discrete jumps, such as excitations or reactive collisions, have to be combined with the continuous change due to flow in phase space. Although it is actually possible to construct a master equation that incorporates both aspects, in this article a simpler approach will be used, based on the moments. The first and second moments vary in time as a consequence of both discrete and continuous processes. It is therefore possible to compute the two kinds of processes separately and compound them in the equations for the moments.

Our interest is focused on the correlation of fluctuations at two points in space or phase space at the same time. We do not consider the correlations at different terms needed to find the fluctuation spectrum. In stationary states they can be derived from our results by using the linearized macroscopic equations together with Onsager's assumption [7].

The equation for the first moments (or average) is the macroscopic equation. The second moments (also called density-density or two-point correlation functions) describe the fluctuations. Equivalent with the second moments, but often more convenient, are the variance and covariance (also called second cumulants). Even more convenient in many cases, however, are the less familiar 'factorial cumulants', whose properties are reviewed in Appendix A.

Our strategy will be, rather than to formulate the method in full generality, to demonstrate it on a series of examples so as to introduce successively various generalizations and complications. Inevitably the examples are somewhat artificial, because most applications to realistic cases require lengthier calculations than would be consistent with the purpose of elucidating the basic idea of the method. Yet the derivation of the fluctuating term inherent in the diffusion equation (Sec.5) and the calculation of the fluctuations in the much discussed chemical reaction in Secs. 8 and 9 may be of practical interest.

2. DECAY WITH DIFFUSION

The aim of this section is to demonstrate the method of compounding moments on the simplest possible example. Take the dissociation of a chemical compound, or the decay of a radio-active isotope

$$X \xrightarrow{a} A, \tag{2.1}$$

where A is inert and does not enter into the equations. Let $N = 0,1,2,\ldots$ be the number of active particles X. The macroscopic rate equation for N as a function of t is of course *

$$d_t N = -aN. \tag{2.2}$$

This is an approximate law, valid when the number N is large. The actual value of N fluctuates around the value given by (2.2) owing to the statistical character of the individual decays. In order to describe the fluctuations one introduces the probability $P(N,t)$ for having N active particles at time t. The evolution of P is governed by the master equation

* For typographical reasons we write d_t for d/dt or $\partial/\partial t$.

$$d_t P(N,t) = a(N+1)\, P(N+1,t) - a N P(N,t), \tag{2.3}$$

which is simply a gain-loss equation for the probability. For its justification one only needs the same assumption on which (2.2) is based: each active particle has a fixed probability a per unit time to decay. Of course this implies the Markov property in that a does not depend on the history of the particle.

The object is to solve (2.3) with arbitrary initial value $P(N,0)$; it clearly suffices to take

$$P(N,0) = \delta_{N,N_o}, \qquad N_o \text{ arbitrary.} \tag{2.4}$$

The solution will determine the average of N, and the fluctuations around it at all $t > 0$. For our purpose it suffices to find the average and the mean square of N.

It is convenient to introduce the step operator E defined by

$$E f(N) = f(N+1),$$

for an arbitrary function f. Note the identity

$$\sum_{N=o}^{\infty} g(N)\, E f(N) = \sum_{N=o}^{\infty} f(N)\, E^{-1} g(N) \tag{2.5}$$

for any pair of functions f,g such that the sum converges and that $g(0) = 0$. With the aid of this step operator the master equation (2.3) may be written

$$d_t P = a\,(E-1)\, N P. \tag{2.6}$$

By definition the master equation is linear in the unknown function P. The term "linear" is therefore available and will now be used to denote the fact that the coefficients in (2.3) and (2.6) are linear functions of N. Linear master equations describe collections of particles that do not interfere with each other — neither by actual forces nor through Bose or Fermi statistics. It will appear that they lead to linear macroscopic laws.

From a mathematical point of view linear master equations have an especially simple structure and can be solved by deriving equations from them for the successive moments. Multiply (2.6) with N and sum over all N, using (2.5)

$$d_t \langle N\rangle = a \sum_{N=o}^{\infty} N(E-1)\, N P = a \langle N(E^{-1}-1) N\rangle = -a \langle N\rangle. \tag{2.7}$$

The angular brackets denote averages. Thus the rate of change of the first moment depends on that same moment alone. In fact, (2.7) is nothing but the macroscopic equation (2.2). Similarly one finds for the second moment

$$d_t \langle N^2\rangle = -2a \langle N^2\rangle + a\langle N\rangle\,. \tag{2.8}$$

Thus its rate of change depends on the first and second moments, but not on higher ones, so that the equations (2.7) and (2.8) can be solved successively.

The solutions corresponding to the initial condition (2.4) are

$$\langle N\rangle_t = N_o\, e^{-at}$$

$$\langle N^2\rangle_t = N_o^2\, e^{-2at} + N_o\, e^{-at}(1-e^{-at})\,. \qquad (2.9)$$

Hence it is possible to obtain the desired information, without determining the complete solution of the master equation. For nonlinear master equations a different approach will be needed (Sec.8).

Although the two moments contain all information we need, two auxiliary quantities will be useful. The variance or second cumulant, to be denoted by double brackets

$$\langle\langle N^2\rangle\rangle = \langle N^2\rangle - \langle N\rangle^2,$$

describes more directly the fluctuations and vanishes for the initial state (2.4). The 'factorial cumulant' will be denoted by square brackets and is defined by (comp. Appendix A)

$$[N^2] = \langle\langle N^2\rangle\rangle - \langle N\rangle = \langle N^2\rangle - \langle N\rangle^2 - \langle N\rangle\,.$$

It often obeys a simpler equation than either the second moment or the variance. For instance in the present case

$$d_t\,\langle\langle N^2\rangle\rangle = -\,2a\,\langle\langle N^2\rangle\rangle + a\,\langle N\rangle$$

$$d_t\,[N^2] = -\,2a\,[N^2]\,.$$

The fact that the latter equation does not have an inhomogeneous term will turn out to greatly simplify the solution in more complicated examples.

So far only the total number N of active particles X has been studied. In order to take into account the location in space, we subdivide the total volume in cells Δ, labelled with a Greek subscript.* The number of particles in each cell λ is N_λ and P is a function of the set $\{N_\lambda\}$. The time dependence will no longer be indicated explicitly. The master equation is

$$d_t\, P(\{N_\lambda\}) = a\sum_\lambda (E_\lambda - 1)\, N_\lambda\, P\,, \qquad (2.10)$$

* This is a purely formal device used as a poor man's substitute for functional integration. It is not the same as the "salami method" (due to Van der Ziel [24,10], also used by Nicolis et al.[19]), since we make no a priori assumptions concerning the correlations between cells.

where E_λ is the step operator acting on the variable N_λ.

The equations for the first two moments are again obtained by multiplying (2.10) with N_α and $N_\alpha N_\beta$ respectively and summing over all $\{N_\lambda\}$,

$$d_t \langle N_\alpha \rangle = - a \langle N_\alpha \rangle \tag{2.11}$$

$$d_t \langle N_\alpha N_\beta \rangle = - 2a \langle N_\alpha N_\beta \rangle + a\delta_{\alpha\beta} \langle N_\alpha \rangle . \tag{2.12}$$

These equations might be surmised directly from (2.7) and (2.8) by noting that there is no statistical correlation among the cells, since so far no transport of particles between them has been included. The solutions can again easily be obtained, but that is not our aim.

The following step is the elimination of the cells by going to a continuous description. Denote the density N_α/Δ in cell α by $n(\vec{r})$, where $\vec{r}$ is the position vector of that cell. Then (2.11) may be written, on dividing by Δ,

$$d_t \langle n(\vec{r}) \rangle = - a \langle n(\vec{r}) \rangle . \tag{2.13}$$

Of course the functions $n(\vec{r})$ and $\langle n(\vec{r}) \rangle$ are only defined on a coarse-grained grid in $\vec{r}$-space, determined by the cell size Δ. In the present example, however, this is not a restriction, because Δ may be taken as small as one wishes. As a check we integrate over an arbitrary volume V to obtain for the number N_V of active particles in V

$$\langle N_V \rangle = \int_V \langle n(\vec{r}) \rangle \, d\vec{r} = - a \langle N_V \rangle .$$

When V is taken to be one of the cells this result reproduces (2.11).

Similarly we divide (2.12) by Δ^2. The Kronecker symbol $\delta_{\alpha\beta}$, considered as a function of β, vanishes everywhere except inside the cell α. Hence its integral over space equals Δ, so that $\delta_{\alpha\beta}/\Delta$ may be identified with Dirac's delta function. Thus (2.12) becomes

$$d_t \langle n(\vec{r}_1)\, n(\vec{r}_2) \rangle = - 2a \langle n(\vec{r}_1)\, n(\vec{r}_2) \rangle + a \langle n(\vec{r}_1) \rangle\, \delta(\vec{r}_1 - \vec{r}_2). \tag{2.14}$$

This can again be justified by integrating $\vec{r}_1$ over cell α and $\vec{r}_2$ over cell β, which reproduces (2.12).

The second cumulant now turns into a mixed cumulant or covariance matrix

$$\langle\langle N_\alpha N_\beta \rangle\rangle = \langle N_\alpha N_\beta \rangle - \langle N_\alpha \rangle \langle N_\beta \rangle ,$$

which obeys

$$d_t \langle\langle N_\alpha N_\beta \rangle\rangle = - 2a \langle\langle N_\alpha N_\beta \rangle\rangle + a\delta_{\alpha\beta} \langle N_\alpha \rangle . \tag{2.15}$$

The factorial cumulant is by definition

$$[N_\alpha N_\beta] = \langle\langle N_\alpha N_\beta \rangle\rangle - \delta_{\alpha\beta} \langle N_\alpha \rangle \tag{2.16}$$

and obeys

$$d_t [N_\alpha N_\beta] = -2a [N_\alpha N_\beta] .$$

Again on dividing by Δ^2, one has in continuous description

$$d_t \langle\langle n(\vec{r}_1)\, n(\vec{r}_2) \rangle\rangle = -2a \langle\langle n(\vec{r}_1)\, n(\vec{r}_2) \rangle\rangle + a\delta(\vec{r}_1 - \vec{r}_2) \langle n(\vec{r}_1) \rangle ,$$

$$d_t [n(\vec{r}_1)\, n(\vec{r}_2)] = -2a\, [n(\vec{r}_1)\, n(\vec{r}_2)] . \tag{2.17}$$

Either of these equations may be used instead of (2.14).

Having obtained the rates at which the first and second moments vary as a result of the decay process, we may now combine them with those caused by transport. Suppose the particles X are molecules in a fluid in which they diffuse with diffusion coefficient D. If only diffusion took place without decay, one would expect

$$d_t \langle n(\vec{r}) \rangle = D \nabla^2 \langle n(\vec{r}) \rangle \tag{2.18}$$

$$d_t \langle n(\vec{r}_1)\, n(\vec{r}_2) \rangle = D(\nabla_1^2 + \nabla_2^2) \langle n(\vec{r}_1)\, n(\vec{r}_2) \rangle \tag{2.19}$$

∇_1^2 and ∇_2^2 are Laplace operators with respect to $\vec{r}_1$ and $\vec{r}_2$. Equation (2.19), however, is not correct, because it treats diffusion as a deterministic process. Actually the diffusion equation (2.18) is the average of a random process. Hence the diffusion is a source of additional noise, which has to be taken into account by adding suitable terms to (2.19). This will be done in Sec.5 and we here anticipate the result: <u>The noisy character of diffusion is correctly taken into account by writing instead of (2.19)</u>

$$d_t [n(\vec{r}_1)\, n(\vec{r}_2)] = (\nabla_1^2 + \nabla_2^2)\, [n(\vec{r}_1)\, n(\vec{r}_2)] . \tag{2.20}$$

Finally we obtain the combined effect of decay and diffusion by adding the effects that each of them has on the moments. First (2.13) and (2.18) lead to

$$d_t \langle n(\vec{r}) \rangle = -a \langle n(\vec{r}) \rangle + D\nabla^2 n(\vec{r}) , \tag{2.21}$$

which is the macroscopic equation. Secondly we compound (2.17) and (2.20) to obtain

$$d_t [n(\vec{r}_1)\, n(\vec{r}_2)] = -2a\, [n(\vec{r}_1)\, n(\vec{r}_2)] + D(\nabla_1^2 + \nabla_2^2)\, [n(\vec{r}_1)\, n(\vec{r}_2)] . \tag{2.22}$$

The equations (2.21) and (2.22) together describe the macroscopic behavior of the density of particles X and the fluctuations around it caused by the statistical character of the individual decay events, and by the stochastic nature of diffusion.

It is easy to solve (2.21) and (2.22) for given initial density $n_o(\vec{r})$. The solution of (2.21) is readily seen to be

$$<n(\vec{r})>_t = \frac{e^{-at}}{(4\pi Dt)^{3/2}} \int \exp\left\{-\frac{(\vec{r}-\vec{r}')^2}{4Dt}\right\} n_o(\vec{r}')\, d\vec{r}'. \quad (2.23)$$

Similarly the general solution of (2.22) is

$$[n(\vec{r}_1)\, n(\vec{r}_2)]_t = \frac{e^{-2at}}{(4\pi Dt)^3} \times$$

$$\times \int \exp\left\{-\frac{(\vec{r}_1-\vec{r}_1')^2+(\vec{r}_2-\vec{r}_2')^2}{4Dt}\right\} [n(\vec{r}_1')\, n(\vec{r}_2')]_o\, d\vec{r}_1'\, d\vec{r}_2' .$$

As the initial value is given by $[n(\vec{r}_1')n(\vec{r}_2')]_o = -\delta(\vec{r}_1-\vec{r}_2)n_o(\vec{r}_1)$, this reduces to

$$[n(\vec{r}_1)\, n(\vec{r}_2)]_t = \frac{-e^{-2at}}{(4\pi Dt)^3} \exp\left\{-\frac{r_1^2+r_2^2}{4Dt}\right\} \times$$

$$\times \int \exp\left\{-\frac{(\vec{r}'^2-(\vec{r}_1+\vec{r}_2)\cdot\vec{r}'}{2Dt}\right\} n_o(\vec{r}')\, d\vec{r}' =$$

$$= \frac{-e^{-2at}}{(4\pi Dt)^3} \exp\left\{-\frac{(\vec{r}_1-\vec{r}_2)^2}{8Dt}\right\} \int \exp\left\{-\frac{\vec{r}'^2}{2Dt}\right\} n_o\left(\vec{r}'+\frac{\vec{r}_1+\vec{r}_2}{2}\right) d\vec{r}' . \quad (2.24)$$

The covariance of the fluctuations $\langle\langle n(\vec{r}_1)\, n(\vec{r}_2)\rangle\rangle$ is obtained by multiplying (2.23) with $\delta(\vec{r}_1-\vec{r}_2)$ and adding it to (2.24). We give the result for the special case $n_o(\vec{r})$ = constant = n_o :

$$\langle\langle n(\vec{r}_1)n(\vec{r}_2)\rangle\rangle_t = \delta(\vec{r}_1-\vec{r}_2)\, n_o\, e^{-at} - \frac{n_o\, e^{-2at}}{(8\pi Dt)^{3/2}} \exp\left\{-\frac{(\vec{r}_1-\vec{r}_2)^2}{8Dt}\right\} .$$

Obviously it only depends on $|\vec{r}_1-\vec{r}_2|$. On integrating over $\vec{r}_1$ and $\vec{r}_2$ one obtains the variance in the total number N, in agreement with (2.9).

3. THE DENSITY IN PHASE SPACE

In this section the method of compounding moments is extended to a distribution in phase space, rather than just coordinate space. As example we consider the same decay process (2.1), but suppose that the particles move freely in space rather than diffusing in a medium. It is then necessary to distinguish them according to both their position $\vec{r}$ and velocity $\vec{v}$. Accordingly the cells Δ are now six-dimensional cells in the one-particle phase space and $n(\vec{r},\vec{v})$ is the phase space density. For chemical reactions it is more realistic to allow a to depend on $|\vec{v}| = v$. The master equation for the cell distribution then takes the form

$$d_t P(\{N_\lambda\}) = \sum_\lambda a_\lambda(E_\lambda - 1) N_\lambda P . \tag{3.1}$$

The equations for the moments are

$$d_t \langle N_\alpha \rangle = - a_\alpha \langle N_\alpha \rangle$$

$$d_t \langle N_\alpha N_\beta \rangle = - (a_\alpha + a_\beta) \langle N_\alpha N_\beta \rangle + \delta_{\alpha\beta}\, a_\alpha \langle N_\alpha \rangle .$$

In the continuous description one now has to put

$$\delta_{\alpha\beta} / \Delta \;\rightarrow\; \delta(\vec{r}_1 - \vec{r}_2)\, \delta(\vec{v}_1 - \vec{v}_2).$$

The result is

$$d_t \langle n(\vec{r},\vec{v}) \rangle = - a(v) \langle n(\vec{r},\vec{v}) \rangle$$

$$d_t \langle n(\vec{r}_1,\vec{v}_1)\, n(\vec{r}_2,\vec{v}_2) \rangle = - \left\{ a(v_1) + a(v_2) \right\} \langle n(\vec{r}_1,\vec{v}_1)\, n(\vec{r}_2,\vec{v}_2) \rangle + \delta(\vec{r}_1 - \vec{r}_2)\, \delta(\vec{v}_1 - \vec{v}_2)\, a(v_1) \langle n(\vec{r}_1,\vec{v}_1) \rangle .$$

The last equation again simplifies when written in terms of the factorial cumulant:

$$d_t [n(\vec{r}_1,\vec{v}_1)\, n(\vec{r}_2,\vec{v}_2)] = - \left\{ a(v_1) + a(v_2) \right\} [n(\vec{r}_1,\vec{v}_1)\, n(\vec{r}_2,\vec{v}_2)] . \tag{3.2}$$

These equations are now compounded with the flow term:

$$d_t \langle n(\vec{r},\vec{v}) \rangle = - \vec{v} \cdot \nabla \langle n(\vec{r},\vec{v}) \rangle - a(\vec{v}) \langle n(\vec{r},\vec{v}) \rangle \tag{3.3}$$

$$d_t \langle n(\vec{r}_1,\vec{v}_1)\, n(\vec{r}_2,\vec{v}_2) \rangle = (- \vec{v}_1 \cdot \nabla_1 - \vec{v}_2 \cdot \nabla_2) \langle n(\vec{r}_1,\vec{v}_1)\, n(\vec{r}_2,\vec{v}_2) \rangle - \left\{ a(v_1) + a(v_2) \right\} \langle n(\vec{r}_1,\vec{v}_1) n(\vec{r}_2,\vec{v}_2) \rangle + \delta(\vec{r}_1 - \vec{r}_2) \delta(\vec{v}_1 - \vec{v}_2) a(\vec{v}_1) \langle n(\vec{r}_1,\vec{v}_1) \rangle . \tag{3.4}$$

In this case the flow terms are to be inserted in the second moment rather than in the factorial cumulant, since the flow in phase space is a deterministic process and not an additional noise source (Appendix B). The equations (3.3) and (3.4) determine the average flow in phase space and the fluctuations around it.

It is again somewhat simpler to rewrite (3.4) in terms of the factorial cumulant. After some algebra one finds the simple result

$$d_t[n(\vec{r}_1,\vec{v}_1)\ n(\vec{r}_2,\vec{v}_2)] =$$
$$= -\left\{\vec{v}_1\cdot\nabla_1 + \vec{v}_2\cdot\nabla_2 + a(v_1) + a(v_2)\right\}[n(\vec{r}_1,\vec{v}_1)\ n(\vec{r}_2,\vec{v}_2)]\ . \quad (3.5)$$

On comparing this with (3.2) one observes that actually no error is made by inserting the flow terms in the equation for the factorial cumulant instead of in the equation for the second moment. This fact is proved generally in Appendix B.

The solution of (3.3) with given initial $n_o(\vec{r},\vec{v})$ is

$$\langle n(\vec{r},\vec{v})\rangle_t = e^{-a(v)t}\, n_o(\vec{r}-\vec{v}t,\vec{v}).$$

Similarly (3.5) is solved by

$$[n(\vec{r}_1,\vec{v}_1)n(\vec{r}_2,\vec{v}_2)]_t = e^{-\{a(v_1)+a(v_2)\}t}\,[n(\vec{r}_1-\vec{v}_1t,\vec{v}_1)\, n(\vec{r}_2-\vec{v}_2t,\vec{v}_2)]_o$$
$$= e^{-2a(v_1)t}\,\delta(\vec{r}_1-\vec{r}_2)\delta(\vec{v}_1-\vec{v}_2)n_o(\vec{r}_1-\vec{v}_1t,\vec{v}_1)$$

Hence the correlation of the fluctuations in phase space are

$$\langle\langle n(\vec{r}_1,\vec{v}_1)\ n(\vec{r}_2,\vec{v}_2)\rangle\rangle_t =$$
$$= \delta(\vec{r}_1-\vec{r}_2)\ \delta(\vec{v}_1-\vec{v}_2)\ e^{-a(v_1)t}\{1-e^{-a(v_1)t}\}\, n_o(\vec{r}_1-\vec{v}_1 t,\vec{v}_1)$$
$$= \delta(\vec{r}_1-\vec{r}_2)\ \delta(\vec{v}_1-\vec{v}_2)\ \{1-e^{-a(v_1)t}\}\langle n(\vec{r}_1,\vec{v}_1)\rangle_t\ .$$

It appears that there is no correlation between the fluctuations at different points of phase space. The reason is again that the particles do not interfere with one another.

4. FLUCTUATIONS IN A STATIONARY STATE

The decay process in the previous two sections tends to an equilibrium in which everything vanishes. In this section we add a term that creates particles, so as to make a nontrivial stationary state possible. The scheme is

$$B \xrightarrow{b} X \qquad X \xrightarrow{a} A\ , \quad (4.1)$$

and may be regarded as an intermediate stage of a radio-active decay process, or as a chemical reaction. B is supposed to be a reservoir of practically constant amount. The probability per unit time that an X is created in a given cell λ is proportional to the amount of B in that cell and will therefore be denoted by $b\Delta$. The master equation describing (4.1) is

$$d_t P(\{N_\lambda\}) = b\Delta \sum_\lambda (E_\lambda^{-1} - 1) P + a \sum_\lambda (E_\lambda - 1) N_\lambda P, \qquad (4.2)$$

where N_λ is again the number of particles X in cell λ.

As this master equation is still linear it is possible to derive from it equations for the separate moments

$$d_t < N_\alpha > = b\Delta - a < N_\alpha >$$

$$d_t < N_\alpha N_\beta > = b\Delta \left\{ < N_\alpha > + < N_\beta > + \delta_{\alpha\beta} \right\} + a \left\{ -2 < N_\alpha N_\beta > + \delta_{\alpha\beta} < N_\alpha > \right\}.$$

In terms of the variance the latter equation simplifies to

$$d_t \ll N_\alpha N_\beta \gg = -2a \ll N_\alpha N_\beta \gg + \delta_{\alpha\beta} \left\{ b\Delta + a < N_\alpha > \right\},$$

and in terms of the factorial cumulant to

$$d_t [N_\alpha N_\beta] = -2a [N_\alpha N_\beta]. \qquad (4.3)$$

Dividing by Δ and Δ^2 respectively and using the continuous notation one gets

$$d_t < n(\vec{r}) > = b - a < n(\vec{r}) >$$

$$d_t [n(\vec{r}_1) n(\vec{r}_2)] = -2a [n(\vec{r}_1) n(\vec{r}_2)].$$

Now suppose again that the reaction occurs in a diffusive medium and add the corresponding terms :

$$d_t < n(\vec{r}) > = b - a < n(\vec{r}) > + D \nabla^2 < n(\vec{r}) > \qquad (4.4)$$

$$d_t [n(\vec{r}_1) n(\vec{r}_2)] = \left\{ -2a + D (\nabla_1^2 + \nabla_2^2) \right\} [n(\vec{r}_1) n(\vec{r}_2)]. \qquad (4.5)$$

These equations are easily solved; the solution of (4.4) is obtained from (2.23) by replacing n with $n - b/a$, while the solution of (4.5) is identical with (2.24). However, we here merely want to study the stationary state given by

$$< n(\vec{r}) >^{st} = b/a$$

$$D(\nabla_1^2 + \nabla_2^2) [n(\vec{r}_1) n(\vec{r}_2)]^{st} = -2a [n(\vec{r}_1) n(\vec{r}_2)]^{st}.$$

It is easily seen that the latter equation has only one solution that does not grow exponentially for large $|\vec{r}_1|$ or $|\vec{r}_2|$, viz.,

$$[n(\vec{r}_1) n(\vec{r}_2)]^{st} = 0.$$

The variance of the fluctuations is therefore

$$\ll n(\vec{r}_1) n(\vec{r}_2) \gg^{st} = \delta(\vec{r}_1 - \vec{r}_2) < n(\vec{r}_1) >. \qquad (4.6)$$

Conclusions: (i) in the stationary state there is no correlation between fluctuations at different points in space ; (ii) the diffusion does not affect the fluctuations ; (iii) the fluctuations are the same as for a Poisson distribution, in agreement with the fact that the individual particles X are independent of each other.

Instead of diffusion we now suppose that the particles move freely in space, as in Sec. 3, and accordingly study the phase space density $n(\vec{r},\vec{v})$. The probability for generating a particle will depend on v, but it will still be supposed independent of $\vec{r}$. In lieu of the two equations (4.4) and (4.5) we now have

$$d_t \langle n(\vec{r},\vec{v})\rangle = b(v) - a(v)\langle n(\vec{r},\vec{v})\rangle - \vec{v}\cdot\nabla \langle n(\vec{r},\vec{v})\rangle$$

$$d_t[n(\vec{r}_1,\vec{v}_1)n(\vec{r}_2,\vec{v}_2)] = -\{a(v_1)+a(v_2)+\vec{v}\cdot\nabla_1+\vec{v}_2\cdot\nabla\}[n(\vec{r}_1,\vec{v}_1)n(\vec{r}_2,\vec{v}_2)]\,.$$

The solutions can again be obtained explicitly, but we are interested in the stationary state

$$\langle n(\vec{r},\vec{v})\rangle^{st} = b(v)\,/\,a(v)$$

$$[n(\vec{r}_1,\vec{v}_1)\; n(\vec{r}_2,\vec{v}_2)]^{st} = 0$$

$$\langle\!\langle n(\vec{r}_1,\vec{v}_1)\; n(\vec{r}_2,\vec{v}_2)\rangle\!\rangle^{st} = \delta(\vec{r}_1-\vec{r}_2)\;\delta(\vec{v}_1-\vec{v}_2)\langle n(\vec{r},\vec{v})\rangle^{st}\,.$$

This result is similar to (4.6) and shows that the propagation does not affect the fluctuations in the stationary state.

5. DIFFUSION AS A STOCHASTIC PROCESS

Let coordinate space be subdivided in cells Δ, and suppose that there is a probability $w_{\lambda\mu}\,dt$ for a particle that is in cell μ at time t to find itself in cell λ at time $t+dt$. The probability distribution $P(\{N_\lambda\})$ of the occupation numbers of the cells obeys the master equation

$$d_t P(\{N_\lambda\}) = \sum_{\lambda\mu} w_{\lambda\mu}\,(E_\lambda^{-1}E_\mu - 1)\,N_\mu\,P\,. \qquad (5.1)$$

The first moment obeys the equation, cf. (2.5),

$$d_t\langle N_\alpha\rangle = \sum_{\lambda\mu} w_{\lambda\mu}\,\langle N_\mu(E_\lambda E_\mu^{-1}-1)\,N_\alpha\rangle$$

$$= \sum_{\lambda\mu} w_{\lambda\mu}\,\langle N_\mu(N_\alpha+\delta_{\alpha\lambda}-\delta_{\alpha\mu}-N_\alpha)\rangle$$

$$= \sum_{\mu} w_{\alpha\mu}\langle N_\mu\rangle - \sum_{\lambda} w_{\lambda\alpha}\langle N_\alpha\rangle\,. \qquad (5.2)$$

To save writing we define the matrix

$$W_{\alpha\beta} = w_{\alpha\beta} - \delta_{\alpha\beta} \sum_{\lambda} w_{\lambda\alpha} \, .$$

Then the equation for the first moments may be written

$$d_t \langle N_\alpha \rangle = \sum_{\mu} W_{\alpha\mu} \langle N_\mu \rangle . \tag{5.3}$$

For the second moments one finds in a similar way

$$d_t \langle N_\alpha N_\beta \rangle = \sum_{\mu} W_{\alpha\mu} \langle N_\mu N_\beta \rangle + \sum_{\mu} W_{\beta\mu} \langle N_\alpha N_\mu \rangle$$

$$- w_{\alpha\beta} \langle N_\beta \rangle - w_{\beta\alpha} \langle N_\alpha \rangle$$

$$+ \delta_{\alpha\beta} \left\{ \sum_{\mu} w_{\alpha\mu} \langle N_\mu \rangle + \sum_{\lambda} w_{\lambda\alpha} \langle N_\alpha \rangle \right\} . \tag{5.4}$$

Again this equation simplifies when written in the factorial cumulant

$$d_t [N_\alpha N_\beta] = \sum_{\mu} W_{\alpha\mu} [N_\mu N_\beta] + \sum_{\mu} W_{\beta\mu} [N_\alpha N_\mu] . \tag{5.5}$$

Writing W for the matrix $W_{\alpha\beta}$ and Θ for the matrix $[N_\alpha N_\beta]$ one has the matrix equation

$$d_t \Theta = W\Theta + \Theta\tilde{W} , \tag{5.6}$$

where the tilde denotes transposition.

In order to go to the continuous description one should realize that $w_{\lambda\mu}$ is proportional to the size of the receiving cell λ and therefore involves a factor Δ. This provides the $d\vec{r}$ needed to change the sums into integrals.

$$w_{\lambda\mu} \rightarrow w(\vec{r}\,|\,\vec{r}\,')\, d\vec{r}$$

$$W_{\lambda\mu} \rightarrow \left\{ w(\vec{r}\,|\,\vec{r}\,') - \delta(\vec{r}-\vec{r}\,') \int d\vec{r}\,'' w(\vec{r}\,''|\vec{r}) \right\} d\vec{r} . \tag{5.7}$$

The equations (5.2) and (5.3) therefore become

$$d_t \langle n(\vec{r}) \rangle = \int w(\vec{r}\,|\,\vec{r}\,') \langle n(\vec{r}\,') \rangle \, d\vec{r}\,' - \int w(\vec{r}\,'|\,\vec{r})\, d\vec{r}\,' . \langle n(\vec{r}) \rangle$$

$$= \int W(\vec{r}\,|\,\vec{r}\,') \langle n(\vec{r}\,') \rangle \, d\vec{r}\,' . \tag{5.8}$$

Similarly (5.5) takes the form

$$d_t[n(\vec{r}_1)\ n(\vec{r}_2)] = \int W(\vec{r}_1|\vec{r}')\ [n(\vec{r}')\ n(\vec{r}_2)]\ d\vec{r}' + \\ + \int W(\vec{r}_2|\vec{r}')\ [n(\vec{r}_1)\ n(\vec{r}')]\ d\vec{r}' , \qquad (5.9)$$

which is also covered by the symbolic form (5.6).

In order that the average density obeys the diffusion equation one must identify the operator W in (5.8) with $D\nabla^2$. It then follows from (5.9) that $[n(\vec{r}_1)\ n(\vec{r}_2)]$ obeys

$$d_t[n(\vec{r}_1)\ n(\vec{r}_2)] = D\ (\nabla_1^2 + \nabla_2^2)\ [n(\vec{r}_1)\ n(\vec{r}_2)] . \qquad (5.10)$$

This is the result anticipated in Sec. 2. Note that (2.19) is indeed incorrect since the second moments obey the more complicated equation (5.4). The equivalent equation in continuous variables is most easily obtained from (5.10):

$$(d_t - D\nabla_1^2 - D\nabla_2^2) < n(\vec{r}_1)\ n(\vec{r}_2) > = 2D\nabla_1 . \nabla_2 \left\{ \delta(\vec{r}_1 - \vec{r}_2) < n(\vec{r}_1) >_t \right\} . \qquad (5.11)$$

In Appendix C this result is shown to agree with the one obtained by Van der Ziel [26] and Van Vliet [25] using the Langevin approach.

Electrical conduction is often described as a diffusion process biased by an external field $\vec{F}$. The macroscopic equation for the electron density then obeys

$$\frac{\partial n}{\partial t} = -\mu\vec{F}.\nabla n + D\nabla^2 n , \qquad (5.12)$$

where μ is the mobility while charge effects have been ignored. The same equations (5.8) and (5.9) remain valid, but the operator W is now to be identified with $-\mu\vec{F}.\nabla + D\nabla^2$. Hence

$$d_t[n(\vec{r}_1)\ n(\vec{r}_2)] = \left\{ -\mu\vec{F}.(\nabla_1 + \nabla_2) + D(\nabla_1^2 + \nabla_2^2) \right\} [n(\vec{r}_1)\ n(\vec{r}_2)] . \qquad (5.13)$$

Rewriting this in terms of the second moment one obtains

$$\left\{ d_t + \mu\vec{F}.(\nabla_1 + \nabla_2) - D(\nabla_1^2 + \nabla_2^2) \right\} < n(\vec{r}_1)\ n(\vec{r}_2) > = \\ = -2D\nabla_1 . \nabla_2\ \delta(\vec{r}_1 - \vec{r}_2) < n(\vec{r}_1) > . \qquad (5.14)$$

The term on the right is identical with the one in (5.11). The fact that it turns out to be unaffected by the field constitutes a justification for considering it as a noise source inherent in the diffusion process. The mathematical reason why the field does not affect the noise is seen on performing the algebra that leads from (5.13) to (5.14): the field term in W involves a first rather than a second derivative.

6. CONDUCTION BETWEEN ELECTRODES

The equations (5.12) and (5.13) for electrical conduction cannot be solved unless one knows the boundary conditions. They are provided by the nature of the surface of the material, in particular the electrodes attached to it. For simplicity we take a long wire with two electrodes at the ends, so that the problem may be treated as one-dimensional. The precise effect of the electrodes depends on the physical properties of the contact. We choose a simple model for it in order to minimize the algebra [27].

The wire, $0 < x < L$, is subdivided in cells of length Δ with occupation numbers N_λ with $\lambda = 0,1,2,\ldots,\Lambda = L/(N+1)$. The probability distribution $P(\{N_\lambda\})$ obeys a master equation whose main term is (5.1), where W has the form given in (5.12). Additional terms arise from the ends, where electrons spill over from the electrodes and also disappear into them. The end at $x=0$ is responsible for two addition terms in the master equation.

$$\Delta \sum_\lambda b_\lambda (E_\lambda^{-1} - 1)\, P + \sum_\lambda a_\lambda (E_\lambda - 1)\, N_\lambda P .$$

They are similar to (4.2), but the coefficients for creation and annihilation now depend on the cell λ, and are practically zero for cells that are more than a few mean free paths from the end. Adding the same term for the other end one obtains the master equation

$$\begin{aligned} d_t P(\{N_\lambda\}) &= \sum_{\lambda\mu} w_{\lambda\mu} (E_\lambda^{-1} E_\mu - 1)\, N_\mu P \\ &+ \Delta \sum_\lambda b_\lambda (E_\lambda^{-1} - 1)\, P + \sum_\lambda a_\lambda (E_\lambda - 1)\, N_\lambda P \\ &+ \Delta \sum_\lambda b_{\Lambda-\lambda} (E_\lambda^{-1} - 1)\, P + \sum_\lambda a_{\Lambda-\lambda} (E_\lambda - 1)\, N_\lambda P . \end{aligned} \tag{6.1}$$

First compute the rate equation for the average:

$$d_t \langle N_\alpha \rangle = \sum_\mu W_{\alpha\mu} \langle N_\mu \rangle + \Delta (b_\alpha + b_{\Lambda-\alpha}) - (a_\alpha + a_{\Lambda-\alpha}) \langle N_\alpha \rangle .$$

In the continuous description

$$\begin{aligned} \frac{\partial \langle n(x) \rangle}{\partial t} &= -\mu F \frac{\partial \langle n(x) \rangle}{\partial x} + D \frac{\partial^2 \langle n(x) \rangle}{\partial x^2} \\ &+ \left\{ b(x) + b(L-x) \right\} - \left\{ a(x) + a(L-x) \right\} \langle n(x) \rangle . \end{aligned} \tag{6.2}$$

In principle this equation can be solved if b(x) and a(x) are known. It is clear, however, that the precise form of these functions cannot be important for the actual problem since they have a very short range. To utilize that fact we let their range go to zero and at the same time increase their magnitude so as not to loose their effect altogether. More precisely we take

$$b(x) = \frac{1}{\epsilon}\, b\,, \qquad a(x) = \frac{1}{\epsilon}\, a \qquad (0<x<\epsilon)\,; \tag{6.3}$$

$$b(x) = 0\,, \qquad a(x) = 0 \qquad (x>\epsilon)\,. \tag{6.4}$$

In the limit $\epsilon \to 0$ the result is that the boundary values of $\langle n(x)\rangle$ are fixed

$$\langle n(0)\rangle = \langle n(L)\rangle = b/a \equiv \rho\,.$$

Thus the effect of our electrodes on the macroscopic diffusion equation is simply to assign fixed boundary values to n(x).

The solution of the equation consisting of the first line of (6.2) with boundary values ρ is a standard problem. The stationary solution is particularly easy: clearly $\langle n(x)\rangle = \rho$ obeys all requirements. That gives for the average electrical current the familiar expression

$$\langle J\rangle = e\mu F\langle n\rangle - eD\,\frac{d\langle n\rangle}{dx} = e\mu F\rho\,.$$

Of course, knowing that the Boltzmann equilibrium $\langle n\rangle = \exp(eFx/kT)$ carries no current, one finds from this $eD = \mu kT$. (This should have come out automatically if we had specified the transition probabilities $w_{\lambda\mu}$ in Sec. 5, but we have taken a shortcut by replacing the matrix W rightaway with the differential operator (5.12).

The fluctuations obey (5.13), or, in one dimension,

$$\frac{\partial}{\partial t}[n(x_1)\,n(x_2)] = \left\{-\mu F\left(\frac{\partial}{\partial x_1}+\frac{\partial}{\partial x_2}\right) + D\left(\frac{\partial^2}{\partial x_1^2}+\frac{\partial^2}{\partial x_2^2}\right)\right\}[n(x_1)\,n(x_2)]\,. \tag{6.5}$$

To find the proper boundary conditions, however, it is necessary to return to the master equation (6.1). One obtains in the usual way, comp. (5.5) and (4.3),

$$d_t[N_\alpha N_\beta] = \sum_\mu W_{\alpha\mu}[N_\mu N_\beta] + \sum_\mu W_{\beta\mu}[N_\alpha N_\mu] - \left(a_\alpha + a_\beta + a_{\Lambda-\alpha} + a_{\Lambda-\beta}\right)[N_\alpha N_\beta]\,.$$

Even without writing this out in the continuous description one sees that the last term has no effect unless at least one of the cells $\alpha, \beta, \Lambda-\alpha, \Lambda-\beta$ is near a boundary. In that case, however, the corresponding coefficient is large of order a/ϵ, so that $[N_\alpha N_\beta]$ goes to zero in a negligibly short time. Hence the limit $\epsilon \to 0$ in (6.3), (6.4) leads to the boundary values

$$[n(0)\, n(x)] = [n(L)\, n(x)] = 0 .$$

Clearly with these boundary conditions the only stationary solution of (6.5) is $[n(x_1)\, n(x_2)] = 0$, so that

$$\langle n(x_1)\, n(x_2)\rangle = \langle n(x_1)\rangle \langle n(x_2)\rangle + \delta(x_1 - x_2)\langle n(x_1)\rangle . \tag{6.6}$$

This result states that the density fluctuations are Poissonian, just as in (4.6). This is not surprising because the electrons are injected and retrieved by the electrodes independently from each other.

Thus we have found the two-point correlations of the density of charge carriers. Unfortunately this is not enough information for finding the current fluctuations. For that purpose one needs to know the velocity of the carriers, so that a description in terms of the phase space density $n(\vec{r},\vec{v})$ is required. Its average obeys the flow equation

$$\left(\frac{\partial}{\partial t} + \vec{v}\cdot\frac{\partial}{\partial\vec{r}} + \vec{F}\cdot\frac{\partial}{\partial\vec{v}}\right)\langle n(\vec{r},\vec{v})\rangle = W\langle n(\vec{r},\vec{v})\rangle ,$$

where W is the same collision operator as in the next section. It would therefore be necessary to tie up this flow in $\vec{r}, \vec{v}$ space with the diffusion equation in $\vec{r}$ space. This task will not be undertaken here [13, 28].

7. THE LORENTZ GAS

Consider a classical gas of N identical and independent molecules, each endowed with a velocity $\vec{v}$ and having a probability $w(\vec{v}'|\vec{v})$ per unit time to change to a velocity $\vec{v}'$ through the collision with a scatterer. At first we suppose that the scatterers are also moving, so that in general $|\vec{v}'| \neq |\vec{v}|$. The macroscopic equation for the density $n(\vec{v})$ in velocity space is

$$\begin{aligned} d_t\, n(\vec{v}) &= \int \left\{ w(\vec{v}|\vec{v}')\, n(\vec{v}') - w(\vec{v}'|\vec{v})\, n(\vec{v}) \right\} d\vec{v}' \\ &= \int W(\vec{v}|\vec{v}')\, n(\vec{v}')\, d\vec{v}' . \end{aligned} \tag{7.1}$$

This model has been employed for describing electrons in a semiconductor. The solution for given initial density $n_0(\vec{v})$ is formally

$$n(\vec{v},t) = e^{Wt}\, n_0(\vec{v}) .$$

As (7.1) has the form of a master equation we may conclude that all solutions tend to an equilibrium, which is the same function $\chi(\vec{v})$ for all, apart from normalization. Taking $\chi(\vec{v})$ normalized according to $\int \chi(\vec{v})\, d\vec{v} = 1$ one has

$$e^{Wt} n_o \rightarrow \chi(\vec{v})\, N = \chi(\vec{v}) \int n_o(\vec{v}')\, d\vec{v}' . \qquad (7.2)$$

Thus e^{Wt} tends to a projection operator.*

In order to describe the fluctuations about the macroscopic behavior we subdivide velocity space in cells Δ, with occupation numbers N_λ and write the master equation for $P(\{N_\lambda\})$. It is the same equation (5.1) as for diffusion in coordinate space. Hence we can skip the various steps and write immediately the equation for the average by substituting $\vec{v}$ for $\vec{r}$ in (5.8):

$$d_t < n(\vec{v}) > = \int W(\vec{v}|\vec{v}') < n(\vec{v}') > d\vec{v}' . \qquad (7.3)$$

This has of course the same form as the macroscopic equation (7.1). Similarly the equation for the fluctuations can be read off from (5.9):

$$d_t [n(\vec{v}_1)\, n(\vec{v}_2)] = \int W(\vec{v}_1|\vec{v}') [n(\vec{v}')\, n(\vec{v}_2)]\, d\vec{v}' +$$

$$+ \int W(\vec{v}_2|\vec{v}') [n(\vec{v}_1)\, n(\vec{v}')]\, d\vec{v}' . \qquad (7.4)$$

In principle this solves the problem of finding the fluctuations in an arbitrary time-dependent situation, but one cannot make the result more explicit without solving (7.3). The fluctuations in equilibrium, however, can readily be found. Equation (7.4) has the form (5.6) and is solved formally by

$$\Theta(t) = e^{Wt}\, \Theta(0)\, e^{\tilde{W}t} , \qquad (7.5)$$

as can be checked by direct substitution. With the aid of (7.2) one obtains for $t \rightarrow \infty$

$$\Theta(\infty) = [n(\vec{v}_1)\, n(\vec{v}_2)]^{eq} = \chi(\vec{v}_1)\, \chi(\vec{v}_2) \iint [n(\vec{v}_1')\, n(\vec{v}_2')]_o\, d\vec{v}_1' d\vec{v}_2'$$

$$= - \chi(\vec{v}_1)\, \chi(\vec{v}_2)\, N .$$

This gives for the covariance of the density

$$\langle\langle n(\vec{v}_1)\, n(\vec{v}_2) \rangle\rangle^{eq} = N \left\{ \delta(\vec{v}_1 - \vec{v}_2)\, \chi(\vec{v}_1) - \chi(\vec{v}_1)\, \chi(\vec{v}_2) \right\}, \qquad (7.6)$$

and for the second moment

$$< n(\vec{v}_1) n(\vec{v}_1) >^{eq} = (1 - N^{-1}) < n(\vec{v}_1) >^{eq} < n(\vec{v}_2) >^{eq} + \delta(\vec{v}_1 - \vec{v}_2) < n(\vec{v}_1) >^{eq} .$$

(7.7)

* The precise definition of a projection operator usually implies that it must be hermitian; it can be seen that (7.2) meets this requirement when a suitable scalar product is defined.

This result agrees with the familiar counting argument of equilibrium statistical mechanics. Subdivide velocity space in cells λ each having an a priori probability p_λ. Distribute the N molecules over these cells; the probability for a set of occupation numbers $\{N_\lambda\}$ with $\Sigma N_\lambda = N$ is

$$\frac{N!}{N_1!\,N_2!\ldots} p_1^{N_1} p_2^{N_2} \ldots = N!\prod_\lambda \frac{p_\lambda^{N_\lambda}}{N_\lambda!} .$$

The average $\langle N_\alpha \rangle$ is $N p_\alpha$ and

$$\begin{aligned}\langle N_\alpha N_\beta \rangle &= N(N-1)\, p_\alpha p_\beta + \delta_{\alpha\beta} N p_\alpha \\ &= (1 - N^{-1})\, \langle N_\alpha\rangle \langle N_\beta\rangle + \delta_{\alpha\beta} \langle N_\alpha \rangle .\end{aligned}$$

in agreement with (7.7). It deviates from the Poisson distribution because of the term involving $1/N$, which is due to the fact that the total number is fixed.

Now modify the model by supposing that the scatterers are fixed. As a consequence no transitions occur with $|\vec{v}| \neq |\vec{v}'|$, so that

$$W(\vec{v}|\vec{v}') = v^{-2}\delta(v - v')\; W_v(\Omega/\Omega') , \tag{7.8}$$

where Ω is the unit vector $\vec{v}/v$. The factor v^{-2} has been inserted for convenience in normalizing, in agreement with

$$\delta(\vec{v} - \vec{v}') = v^{-2}\,\delta(v - v')\,\delta(\Omega - \Omega') . \tag{7.9}$$

We set $N_v\,dv$ for the number of molecules between v and $v + dv$,

$$N_v = \int \delta(v - |\vec{v}'|)\, n(\vec{v}')\, d\vec{v}' .$$

Equation (7.1) no longer has a unique equilibrium solution, because the collisions are unable to alter N_v. Assuming that there are no other constants of the motion, one has an equilibrium distribution $\chi(v, \Omega)$ for each separate v, which is of course independent of the direction Ω, and properly normalized equals $1/4\pi$. Hence in the limit

$$e^{Wt}\, n_0(\vec{v}) \;\rightarrow\; \frac{N_v}{4\pi} = \frac{1}{4\pi}\int \delta(v - v')\, n_0(\vec{v}')\, d\vec{v}' ,$$

which is again a projection operator. Thus $\langle n(\vec{v}')\rangle_t$ tends to $\langle n(\vec{v})\rangle_\infty = N_v/4\pi$.

The solution (7.5) for Θ yields in the limit

$$\begin{aligned}[n(\vec{v}_1)n(\vec{v}_2)]_\infty &= -(4\pi)^{-2}\int \delta(v_1 - v_1')d\vec{v}_1'\int \delta(v_2 - v_2')d\vec{v}_2'\; \delta(\vec{v}_1' - \vec{v}_2')n_0(\vec{v}_2')d\vec{v}_2' \\ &= -(4\pi)^{-2}\;\delta(v_1 - v_2)\, N_{v_1} .\end{aligned}$$

Hence, using (7.9),

$$\langle\!\langle n(\vec{v}_1)n(\vec{v}_2)\rangle\!\rangle_\infty = -(4\pi)^{-2}\,\delta(v_1-v_2)\,N_{v_1} + (4\pi)^{-1}\,\delta(\vec{v}_1-\vec{v}_2)\,N_{v_1}$$
$$= \delta(v_1-v_2)\,N_{v_1}\left\{\frac{1}{4\pi v_1^2}\,\delta(\Omega_1-\Omega_2) - \frac{1}{(4\pi)^2}\right\}.$$

This is analogous to (7.6), but has a delta function in front, which states that no correlations exist between subspaces belonging to different v. The second moment can also be written in a form analogous to (7.7)

$$<n(\vec{v}_1)\;n(\vec{v}_2)>_\infty =$$
$$= \left\{1-\frac{\delta(v_1-v_2)}{N_{v_1}}\right\}<n(\vec{v}_1)>_\infty\;<n(\vec{v}_2)>_\infty + \delta(\vec{v}_1-\vec{v}_2)<n(\vec{v}_1)>_\infty.$$

8. THE EXPANSION OF A NONLINEAR MASTER EQUATION

So far the examples were all governed by linear master equations, from which closed equations could be obtained for the first and second moments. As a first example of a nonlinear master equation we take the simple chemical reaction

$$B \overset{b}{\rightarrow} X, \qquad 2X \overset{a}{\rightarrow} A\,. \tag{8.1}$$

The compound A may well be the molecule X_2, provided that it does not dissociate again. This model has been the subject of some discussion[19-22]. In the homogeneous stationary state the fluctuations in the total number of molecules X can be calculated without difficulty[5], but we are now in a position to compute the spatial correlations between local fluctuations as well.

Subdivide the total volume into cells Δ, with occupation numbers N_λ. The master equation for the probability distribution $P(\{N_\lambda\})$ is, ignoring transport,

$$d_t\,P(\{N_\lambda\}) = b\Delta\sum_\lambda (E_\lambda^{-1}-1)\,P + a\Delta^{-1}\sum_\lambda (E_\lambda^2-1)\,N_\lambda^2\,P\,. \tag{8.2}$$

The first term is the same as in (4.2). The second term expresses that each of the N_λ molecules X in cell λ has a probability proportional to N_λ/Δ to meet another one in unit time. Actually the number of pairs is of course $\frac{1}{2}N_\lambda(N_\lambda-1)$, but the 1 is immaterial in the order that we shall calculate[21,5], and the factor $\frac{1}{2}$ has been absorbed in the reaction rate a, as is customary when using the law of mass action.

It is no longer possible to deduce from (8.2) exact closed equations for the first and second moments, but an expansion method exists, based on the idea that N_λ is large [1-4]. Note that it is therefore no longer possible to take Δ arbitrarily small. Nevertheless the language of continuously varying variables can be employed, provided one does not claim to describe fluctuations on a scale comparable with the mean distance between molecules X. We shall now demonstrate this expansion method on (8.2).

Transform from the variable N_λ to new variables ξ_λ by setting

$$N_\lambda = \Delta\varphi_\lambda(t) + \Delta^{\frac{1}{2}}\xi_\lambda , \tag{8.3}$$

where the φ_λ are functions of time to be specified presently. This substitution decomposes N_λ into a "macroscopic part" proportional to Δ and a "fluctuating part" proportional to $\Delta^{\frac{1}{2}}$. In chemical parlance φ_λ is the concentration.

The distribution function $P(\{N_\lambda\},t)$ transforms into a distribution $\Pi(\{\xi_\lambda\},t)$,

$$P(\{\Delta\varphi_\lambda(t) + \Delta^{\frac{1}{2}}\xi_\lambda\}, t) = \Delta^{-\frac{1}{2}\Lambda}\,\Pi(\{\xi_\lambda\},t).$$

The factor $\Delta^{-\frac{1}{2}\Lambda}$, where Λ is the total number of cells, is needed for the normalization of Π but will not enter into the equations. The operator E_λ acting on an arbitrary function $f(\xi_\lambda)$ takes the form

$$E_\lambda f(\xi_\lambda) = f(\xi_\lambda + \Delta^{-\frac{1}{2}}) = \left\{1 + \Delta^{-\frac{1}{2}}\frac{\partial}{\partial\xi_\lambda} + \tfrac{1}{2}\Delta^{-1}\frac{\partial}{\partial\xi_\lambda^2} + \dots\right\} f(\xi_\lambda).$$

Rewriting the master equation (8.2) in the new variables one obtains

$$\begin{aligned}\frac{\partial\Pi}{\partial t} - \Delta^{\frac{1}{2}}\sum_\lambda \frac{d\varphi_\lambda}{dt}\frac{\partial\Pi}{\partial\xi_\lambda} &= b\Delta^{\frac{1}{2}}\sum_\lambda\left\{-\frac{\partial}{\partial\xi_\lambda} + \tfrac{1}{2}\Delta^{-\frac{1}{2}}\frac{\partial^2}{\partial\xi_\lambda^2}\right\}\Pi \\ &\quad + a\Delta^{\frac{1}{2}}\sum_\lambda\left\{2\frac{\partial}{\partial\xi_\lambda} + 2\Delta^{-\frac{1}{2}}\frac{\partial^2}{\partial\xi_\lambda^2}\right\}\left(\varphi_\lambda + \Delta^{-\frac{1}{2}}\xi_\lambda\right)^2\Pi .\end{aligned} \tag{8.4}$$

It is now possible to separate the successive orders in Δ. Terms of higher order in $\Delta^{-\frac{1}{2}}$ than we are interested in have already been omitted in (8.4).

The terms of order $\Delta^{\frac{1}{2}}$ are all proportional to $\partial\Pi/\partial\xi_\lambda$ and can therefore be caused to cancel by setting (for each λ)

$$-d_t\varphi_\lambda = -b + 2a\varphi_\lambda^2 . \tag{8.5}$$

These equations determine the behavior of the macroscopic parts of the N_λ and, in fact, they are the familiar rate equations for (8.1). Remember that transfer between cells is still ignored.

The terms of order Δ^o in (8.4) are

$$\frac{\partial \Pi}{\partial t} = \tfrac{1}{2} b \sum_\lambda \frac{\partial^2 \Pi}{\partial \xi_\lambda^2} + 2a \sum_\lambda \varphi_\lambda^2 \frac{\partial^2 \Pi}{\partial \xi_\lambda^2} + 4a \sum_\lambda \varphi_\lambda \frac{\partial}{\partial \xi_\lambda} \xi_\lambda \Pi . \qquad (8.6)$$

This is a multivariate Fokker-Planck equation with coefficients that are linear in the ξ_λ but will depend on time through the φ_λ. It can readily be solved*, but all we need are the first and second moments of the ξ_λ. One easily deduces

$$d_t < \xi_\alpha > \ = -4a\, \varphi_\alpha(t) < \xi_\alpha > \qquad (8.7)$$

$$d_t < \xi_\alpha \xi_\beta > \ = -4a\,(\varphi_\alpha + \varphi_\beta) < \xi_\alpha \xi_\beta > + \delta_{\alpha\beta}\,(b + 4a\varphi_\alpha^2) . \qquad (8.8)$$

In (8.6) — and therefore also in (8.7) and (8.8) — terms of relative order $\Delta^{-\frac{1}{2}}$ are neglected. Note that (8.7) is the "variational equation" associated with (8.5).

Substituting the results (8.5) and (8.7) into (8.3) one obtains

$$d_t < N_\alpha > \ = b\Delta - 2a\Delta^{-1} < N_\alpha >^2 + \Theta(\Delta^o) . \qquad (8.9)$$

In addition one obtains from (8.8) an equation, which is more easily written in terms of the covariance matrix,

$$d_t \ll N_\alpha N_\beta \gg \ = -4a\Delta^{-1}(< N_\alpha > + < N_\beta >) \ll N_\alpha N_\beta \gg +$$

$$+ \delta_{\alpha\beta} \left\{ b\Delta + 4a\Delta^{-1} < N_\alpha >^2 \right\} + \Theta(\Delta^{\frac{1}{2}}) .$$

In order to write these equations with continuous space variables so as to eliminate the cell size Δ we divide them by Δ and Δ^2 respectively and obtain

$$d_t < n(\vec{r}) > \ = \ b - 2a < n(\vec{r}) >^2 \qquad (8.10)$$

$$d_t \ll n(\vec{r}_1)\, n(\vec{r}_2) \gg \ = -4a\,\{ < n(\vec{r}_1) > + < n(\vec{r}_2) > \} \ll n(\vec{r}_1)\, n(\vec{r}_2) \gg$$

$$+ \ \delta(\vec{r}_1 - \vec{r}_2)\ \{ b + 4a < n(\vec{r}_1) >^2 \} . \qquad (8.11)$$

The equation for the factorial cumulant is

$$d_t [n(\vec{r}_1) n(\vec{r}_2)] = -4a\,\{ < n(\vec{r}_1) > + < n(\vec{r}_2) > \}\ [n(\vec{r}_1) n(\vec{r}_2)] -$$

$$- 2a\,\delta(\vec{r}_1 - \vec{r}_2) < n(\vec{r}_1) >^2 . \qquad (8.12)$$

It is clear that the use of the factorial cumulant no longer gives a

* The solution of (8.6) is a Gaussian with mean and variance determined by (8.7) and (8.8).

worthwhile simplification. Yet (8.12) is needed for our next task: compounding these equations with the flow.

9. THE SAME REACTION WITH DIFFUSION

It is now possible to add the effect of transfer between cells. We take the case of transfer by diffusion. The equations (8.10) and (8.12) are now extended into

$$d_t \langle n(\vec{r})\rangle = b - 2a \langle n(\vec{r})\rangle^2 + D\nabla^2 \langle n(\vec{r})\rangle \tag{9.1}$$

$$d_t [n(\vec{r}_1)n(\vec{r}_2)] = \left\{-4a\left(\langle n(\vec{r}_1)\rangle + \langle n(\vec{r}_2)\rangle\right) + D(\nabla_1^2 + \nabla_2^2)\right\} \times$$
$$\times [n(\vec{r}_1)n(\vec{r}_2)] - 2a\,\delta(\vec{r}_1 - \vec{r}_2)\langle n(\vec{r}_1)\rangle^2 . \tag{9.2}$$

As it does not seem possible to obtain explicit solutions of these time dependent equations we apply them to the homogeneous stationary case. First one gets from (9.1)

$$\langle n\rangle^{st} = \sqrt{b/2a},$$

to that (9.2) becomes

$$D(\nabla_1^2 + \nabla_2^2)[n(\vec{r}_1)n(\vec{r}_2)] = 4\sqrt{2ab}\,[n(\vec{r}_1)n(\vec{r}_2)] + b\,\delta(\vec{r}_1 - \vec{r}_2) . \tag{9.3}$$

A particular solution can be obtained by taking for $[n(r_1)n(r_2)]$ a function ϑ of $\vec{r} = \vec{r}_1 - \vec{r}_2$ alone, which then must satisfy

$$\nabla^2 \vartheta(\vec{r}) = \kappa^2 \vartheta(\vec{r}) + (b/2D)\,\delta(\vec{r}),$$

where $\kappa^2 = 2\sqrt{2ab}/D$. Consequently

$$\vartheta(\vec{r}) = -\frac{b}{8\pi D}\frac{e^{-\kappa r}}{r} \tag{9.4}$$

is a solution of (9.3). Other solutions differ from it by a solution of the homogeneous equation and it is easily seen that they have to be excluded since they grow exponentially with $|\vec{r}_1|$ and $|\vec{r}_2|$.

Thus the fluctuations in the homogeneous stationary state are determined by (9.4), or in the more familiar terms of covariance

$$\langle\!\langle n(\vec{r}_1)n(\vec{r}_2)\rangle\!\rangle = -\frac{b}{8\pi D}\frac{e^{-\kappa|\vec{r}_1 - \vec{r}_2|}}{|\vec{r}_1 - \vec{r}_2|} + \sqrt{\frac{b}{2a}}\,\delta(\vec{r}_1 - \vec{r}_2)$$

$$= \langle n\rangle^{st}\left\{\delta(\vec{r}_1 - \vec{r}_2) - \frac{1}{4}\frac{\kappa^2 e^{-\kappa|\vec{r}_1 - \vec{r}_2|}}{4\pi|\vec{r}_1 - \vec{r}_2|}\right\} . \tag{9.5}$$

The rate at which a given molecule X disappears into A is seen from

(8.9) to be $2a\langle n\rangle^{st} = \sqrt{2ab}$. Hence κ^{-1} measures the distance over which it diffuses before decaying. We now apply (9.5) to two limiting cases.

Take a region V whose linear dimensions are much larger than κ^{-1}. The variance of the number N_V of molecules X in that region is

$$\langle\langle N_V^2 \rangle\rangle^{st} = \iint_{V\,V} \langle\langle n(r_1)\, n(r_2) \rangle\rangle \; dr_1 dr_2 \tag{9.6}$$
$$= \langle n\rangle^{st} \left\{ 1 - \frac{1}{4} \right\} V = \frac{3}{4} \langle N_V\rangle^{st}.$$

This is the same result as obtained by Nitzan and Ross [21], and by Kuramoto [20], although they did not use a systematic expansion. The effect of diffusion has disappeared, as was to be expected since most of the molecules produced in V have no time to diffuse away. The fact that the variance is less than that of a Poisson distribution can be understood on the ground that they are not independent but are annihilated in pairs [5].

On the other hand, when V is small compared to κ^{-1} the diffusion is important, but its effect on the fluctuations depends on the precise shape of V. Yet an estimate of the integral (9.6) can be obtained

$$\langle n\rangle^{st} \left\{ V - \frac{\kappa^2}{16\pi} \frac{V^2}{\ell} \right\} = \langle N_V\rangle^{st} \left\{ 1 - \frac{a}{4\pi\ell} \langle N_V\rangle^{st} \right\},$$

where ℓ is of the order of the diameter of V. It is seen that when the size of V is reduced this tends to $\langle N_V\rangle^{st}$, in agreement with the Poisson distribution. In this situation the fluctuations are mainly determined by the diffusion, so that the interaction between the molecules becomes irrelevant.

This distinction between large and small V was pointed out by Kuramoto [22], albeit that he used D as a parameter to separate both cases. Also he expressed his results in the Fourier transform of γ, for which we find from (9.5)

$$\int \langle\langle n(\vec{r}_1)\, n(\vec{r}_1 + \vec{r}) \rangle\rangle^{st} \, e^{i\vec{q}\cdot\vec{r}} \, d\vec{r} = \langle n\rangle^{st} \, \frac{3 + 4q^2/\kappa^2}{4 + 4q^2/\kappa^2} .$$

This is identical with equation (4) in reference 22.

10. A POPULATION PROBLEM

The Malthus-Verhulst equation for a population of N individuals is [29]

$$d_t N = bN - aN - cN^2/\Omega . \tag{10.1}$$

b is the natural birth rate per individual, a the death rate, and cN/Ω the additional death rate due to overcrowding, which is taken to be proportional to the density N/Ω of the population present, Ω being the size of habitat. This is of course the macroscopic equation, which does not account for the fluctuations caused by the statistical nature of the individual birth and death events. They are described by the master equation for this problem, which can easily be constructed without any other assumptions than those used in arriving at (10.1).

So far the total population N was considered. We now want to include the effect of migration and therefore have to divide Ω in cells Δ with their own population numbers N_λ. The resulting multivariate master equation is easily seen to be

$$d_t\, P(\{N_\lambda\}) = b\sum_\lambda (E_\lambda^{-1} - 1)\, N_\lambda P + \sum_\lambda (E_\lambda - 1)\left\{a N_\lambda + c N_\lambda^2/\Delta\right\} P. \tag{10.2}$$

We proceed to deduce the equations for the first and second moments, which will enable us to add the migration terms.

As (10.2) is nonlinear the transformation (8.3) has to be carried out. The result is a master equation for $\Pi(\{\xi_\lambda\},t)$ similar to (8.4)

$$\frac{\partial\Pi}{\partial t} - \Delta^{\frac{1}{2}}\sum_\lambda \frac{d\varphi_\lambda}{dt}\frac{\partial\Pi}{\partial\xi_\lambda} = b\Delta\sum_\lambda\left\{-\Delta^{-\frac{1}{2}}\frac{\partial}{\partial\xi_\lambda} + \tfrac{1}{2}\Delta^{-1}\frac{\partial^2}{\partial\xi_\lambda^2}\right\}\left\{\varphi_\lambda + \Delta^{-\frac{1}{2}}\xi_\lambda\right\}\Pi$$

$$+\,\Delta\sum_\lambda\left\{\Delta^{-\frac{1}{2}}\frac{\partial}{\partial\xi_\lambda} + \tfrac{1}{2}\Delta^{-1}\frac{\partial^2}{\partial\xi_\lambda^2}\right\}\left\{a\varphi_\lambda + c\varphi_\lambda^2 + \Delta^{-\frac{1}{2}}(a + 2c\varphi_\lambda)\,\xi_\lambda\right\}\Pi\,.$$

The terms proportional to $\Delta^{\frac{1}{2}}$ are caused to cancel by choosing for the φ_λ solutions of

$$-\,d_t\,\varphi_\lambda = -\,b\varphi_\lambda + a\varphi_\lambda + c\varphi_\lambda^2\,. \tag{10.3}$$

The terms proportional to Δ^0 give rise to a linear, multivariate, time-dependent Fokker-Planck equation for Π,

$$\frac{\partial\Pi}{\partial t} = \sum_\lambda (-b + a + 2c\varphi_\lambda)\frac{\partial}{\partial\xi_\lambda}\xi_\lambda\Pi + \tfrac{1}{2}\sum_\lambda (b\varphi_\lambda + a\varphi_\lambda + c\varphi_\lambda^2)\frac{\partial^2\Pi}{\partial\xi_\lambda^2}. \tag{10.4}$$

The equations for the averages are

$$d_t\langle\xi_\lambda\rangle = (b - a - 2c\varphi_\lambda)\langle\xi_\lambda\rangle\,. \tag{10.5}$$

They are the variational equations associated with (10.3). Together with (10.3) they give

$$d_t\langle N_\lambda\rangle = (b - a)\langle N_\lambda\rangle - \frac{c}{\Delta}\langle N_\lambda\rangle^2\,.$$

This is simply the macroscopic equation (10.1) for each separate cell. In the continuous description they take the form

$$d_t\langle n(\vec r)\rangle = (b - a)\langle n(\vec r)\rangle - c\langle n(\vec r)\rangle^2\,.$$

The equation for the second moments obtained from (10.4) is

$$d_t < \xi_\alpha \xi_\beta > = 2(b - a - c\varphi_\alpha - c\varphi_\beta) < \xi_\alpha \xi_\beta > + \delta_{\alpha\beta} (b\varphi_\alpha + a\varphi_\alpha - c\varphi_\alpha^2) .$$

The same equation is obeyed by the covariance matrix $\ll \xi_\alpha \xi_\beta \gg$ and one subsequently finds

$$d_t \ll N_\alpha N_\beta \gg \; = \; 2 \left\{ b - a - c < N_\alpha >/\Delta - c < N_\beta >/\Delta \right\} \ll N_\alpha N_\beta \gg$$

$$+ \; \delta_{\alpha\beta} \, \varphi_\alpha \, (b + a + c\varphi_\alpha) .$$

In the continuous description

$$d_t \ll n(\vec{r}_1) n(\vec{r}_2) \gg \; = 2 \left\{ b - a - c < n(\vec{r}_1) > - c < n(\vec{r}_2) > \right\} \ll n(\vec{r}_1) n(\vec{r}_2) \gg$$

$$+ \; \delta(\vec{r}_1 - \vec{r}_2) < n(\vec{r}_1) > \left\{ b + a + c < n(\vec{r}_1) > \right\} . \qquad (10.6)$$

We also need the equation for the factorial cumulants,

$$d_t [n(\vec{r}_1) n(\vec{r}_2)] \; = 2 \left\{ b - a - c < n(\vec{r}_1) - c < n(\vec{r}_2) > \right\} [n(\vec{r}_1) n(\vec{r}_2)]$$

$$+ \; \delta(\vec{r}_1 - \vec{r}_2) < n(\vec{r}_1) > \left\{ 2b - 2c < n(\vec{r}_1) > \right\} .$$

After this preliminary work it is now possible to add the migration terms. We suppose that the migration occurs by diffusion and write accordingly for the total rate of change of the average and the factorial cumulant

$$d_t < n(\vec{r}) > = (b - a) < n(\vec{r}) > - c < n(\vec{r}) >^2 + D\nabla^2 < n(\vec{r}) > . \qquad (10.7)$$

$$d_t [n(\vec{r}_1) n(\vec{r}_2)] \; = 2 \left\{ b - a - c < n(\vec{r}_1) > - c < n(\vec{r}_2) > \right\} [n(\vec{r}_1) n(\vec{r}_2)]$$

$$+ \; 2\delta(\vec{r}_1 - \vec{r}_2) < n(\vec{r}_1) > \left\{ b - c < n(\vec{r}_1) > \right\} + D(\nabla_1^2 + \nabla_2^2) \; [n(\vec{r}_1) n(\vec{r}_2)] . \qquad (10.8)$$

The solution of these equations contains all information concerning the local average density of the population and the fluctuations about this average. Unfortunately they cannot be solved analytically and we therefore restrict ourselves to computing the fluctuations in the stationary state.

Suppose that the birth rate b is larger than the natural death rate a; then (10.7) has the stationary solution

$$< n(\vec{r}) >^{st} \; = \; \frac{b - a}{c} .$$

Substitution in (10.8) yields

$$\left\{ D(\nabla_1^2 + \nabla_2^2) - 2(b - a) \right\} [n(\vec{r}_1) n(\vec{r}_2)]^{st} = -2(b - a)\,(a/c)\, \delta(\vec{r}_1 - \vec{r}_2) .$$

Observe that this equation has the same form as (9.3). Its solution is, in analogy with (9.4),

$$[n(\vec{r}_1)n(\vec{r}_2)]^{st} = \frac{a(b-a)}{4\pi c\, D} \frac{e^{-\kappa|\vec{r}_1 - \vec{r}_2|}}{|\vec{r}_1 - \vec{r}_2|} ,$$

where $\kappa^2(b-a)/D$. Consequently

$$\langle\langle n(\vec{r}_1)n(\vec{r}_2)\rangle\rangle^{st} = \langle n\rangle^{st} \left\{\delta(\vec{r}_1 - \vec{r}_2) + \frac{a}{4\pi D} \frac{e^{-\kappa|\vec{r}_1 - \vec{r}_2|}}{|\vec{r}_1 - \vec{r}_2|}\right\} .$$

Note that the fluctuations are larger than for a Poisson distribution. Integration over a volume V with diameter large compared to κ^{-1} yields

$$\langle\langle N_V^2\rangle\rangle^{st} = \langle N_V\rangle^{st} \frac{b}{b-a} = \frac{b}{c} .$$

As another interesting example Nicolis et al. [19] studied the Lotka-Volterra model for two interacting populations. They endowed each species with some internal "fitness parameter", which varies continuously. In their calculations they made use of various approximations, and it would be of interest to know whether the present systematic method leads to the same results. Unfortunately the calculations are too lengthy to be included in this article.

APPENDIX A: FACTORIAL CUMULANTS

A probability distribution over the integers N = 0,1,2,... whose ordinary moments $\langle N^r\rangle$ exists, also has cumulants $\langle\langle N^r\rangle\rangle$ and factorial moments

$$\langle N(N-1)(N-2)\ \dots\ (N-r+1)\rangle , \tag{A.1}$$

which are certain combinations of the moments. The <u>factorial cumulants</u> [30] $\Theta^{(r)} \equiv [N^r]$ are obtained by writing out (A.1) in the moments $\langle N^{r-s}\rangle$ and replacing each moment by the corresponding cumulant $\langle\langle N^{r-s}\rangle\rangle$. They are generated by the logarithm of the generating function H(v) of the factorial moments

$$\log H(v) = \log \langle (1+v)^N\rangle = \sum_{r=1}^{\infty} \frac{v^r}{r!}\, \Theta^{(r)} .$$

H(v) is related to the characteristic function G(t) by

$$H(e^{it} - 1) = G(t) . \tag{A.2}$$

The first factorial cumulant is the same as the first moment :

$\Theta^{(1)} = \langle N \rangle$; the Poisson distribution is characterized by the vanishing of all higher $\Theta^{(r)}$.

The generalization to multivariate distribution is trivial, but requires a more elaborate notation.

$$\log \langle \prod_\lambda (1+v_\lambda)^{N_\lambda} \rangle = \sum_{\{r_\lambda\}}{}' \frac{v_1^{r_1} v_2^{r_2} \cdots}{r_1!\, r_2! \cdots} \Theta^{(r_1, r_2, \ldots)} . \qquad \text{(A.3)}$$

The prime indicates that the term with all r_λ equal to zero is absent. Alternatively the right hand side may be written

$$\sum_\lambda v_\lambda \Theta_\lambda + \frac{1}{2} \sum_{\lambda\mu} v_\lambda v_\mu \Theta_{\lambda\mu} + \frac{1}{3!} \sum v_\lambda v_\mu v_\kappa \Theta_{\lambda\mu\kappa} + \ldots ,$$

where for instance, in agreement with (2.16),

$$\Theta_{\lambda\mu} \equiv [N_\lambda N_\mu] = \langle N_\lambda N_\mu \rangle - \langle N_\lambda \rangle \langle N_\mu \rangle - \delta_{\lambda\mu} \langle N_\lambda \rangle .$$

Note that $\Theta_{\lambda\mu} = 0$ when N_λ and N_μ are uncorrelated, and in general all factoral cumulants involving groups of statistically independent variables vanish, just as ordinary cumulants. The factorial cumulant of the sum of two independent variables is the sum of their factorial cumulants ; owing to (A.2) this follows immediately from the corresponding theorem for ordinary cumulants.

The relevance for the present work derives from the following property. The most general <u>linear</u> multivariate master equation contains four types of terms.

(i) Particles λ are created in singles with probability b_λ, or in twins, $b_{\lambda\mu}$, etc. The corresponding term is

$$\left\{ \sum_\lambda b_\lambda (E_\lambda^{-1} - 1) + \sum_{\lambda\mu} b_{\lambda\mu} (E_\lambda^{-1} E_\mu^{-1} - 1) + \ldots \right\} P .$$

(ii) Particles are annihilated with probability a_λ per particle per unit time :

$$\sum_\lambda a_\lambda (E_\lambda - 1) N_\lambda P .$$

(iii) Particles are transferred from μ to λ:

$$\sum_{\lambda\mu} w_{\lambda\mu} (E_\lambda^{-1} E_\mu - 1) N_\mu P .$$

(iv) Particles are created auto-catalytically (or born in litters) :

$$\left\{ \sum_{\lambda\mu\kappa} c^{\lambda}_{\mu\kappa} \left(E^{-1}_{\mu} E^{-1}_{\kappa} E_{\lambda} - 1\right) N_{\lambda} + \sum_{\lambda\mu\kappa\rho} c^{\lambda}_{\mu\kappa\rho} \left(E^{-1}_{\mu} E^{-1}_{\kappa} E^{-1}_{\rho} E_{\lambda} - 1\right) N_{\lambda} + \ldots \right\} P.$$

The solution of the master equation

$$d_t \, P(\{N_\lambda\}) = (i) + (ii) + (iii) + (iv) \qquad (A.4)$$

can be done exactly, but is facilitated by the use of the factorial cumulants.

From (A.4) an equation for factorial moment generating function $H(\{v_\lambda\})$ may be derived. The term (i) yields

$$d_t H = \left\{ \sum_{\lambda} b_\lambda v_\lambda + \sum_{\lambda\mu} b_{\lambda\mu} \left(v_\lambda + v_\mu + v_\lambda + v_\mu\right) + \ldots \right\} H.$$

Divide by H and expand the left-hand side in powers of the v. The result is a set of uncoupled equations for the several factorial cumulants, for instance

$$d_t \Theta_\alpha = b_\alpha + \sum_\mu b_{\alpha\mu} + \sum_\lambda b_{\lambda\alpha} + \sum_{\lambda\mu} \left(b_{\alpha\lambda\mu} + b_{\lambda\alpha\mu} + b_{\lambda\mu\alpha}\right) + \ldots$$

$$d_t \Theta_{\alpha\beta} = b_{\alpha\beta} + b_{\beta\alpha} + \ldots$$

The birth term added in Sec. 4 only involves a b_α but no $b_{\alpha\beta}$; that is the reason why in that section $\Theta_{\alpha\beta}$ obeys the same equation as in Sec. 2.

The term (ii) in (A.4) adds the contribution

$$d_t H = \sum_\lambda a_\lambda \left(\frac{1}{1 + v_\lambda} - 1 \right) (1 + v_\lambda) \frac{\partial H}{\partial v_\lambda} = - \sum_\lambda a_\lambda v_\lambda \frac{\partial H}{\partial v_\lambda} .$$

It affects all factorial cumulants, but in a simple way:

$$d_t \, \Theta_{\alpha\beta\gamma\ldots} = - \left(a_\alpha + a_\beta + a_\gamma + \ldots\right) \Theta_{\alpha\beta\gamma\ldots} \; .$$

This is the reason why in Sec. 2 the factorial cumulant $\Theta_{\alpha\beta}$ obeyed a homogeneous equation. Similarly term (iii) yields

$$d_t H = \sum_{\lambda\mu} w_{\lambda\mu} \left(v_\lambda - v_\mu\right) \frac{\partial H}{\partial v_\mu}$$

and couples the factorial cumulants of the same order with each other. This is the reason why in Sec. 5 the factorial cumulant matrix Θ obeyed the homogeneous equation (5.6).

Finally a typical contribution of (iv) to $d_t H$ is

$$\sum c^{\lambda}_{\mu\kappa\rho} \left(v_\mu + v_\kappa + v_\rho + v_\mu v_\kappa + v_\kappa v_\rho + v_\rho v_\mu + v_\mu v_\kappa v_\rho - v_\lambda\right) \frac{\partial H}{\partial v_\lambda} ,$$

which couples each factorial cumulant to others of the same and lower orders. However, the same can be said about the second moments or the covariances, so that at this stage the factorial moment ceases to have a clear-cut advantage. As to nonlinear master equations, it is pointed out in Sec. 8 that for them the use of the factorial cumulant no longer gives a worthwhile simplification. Yet it is convenient to use for the purpose of adding the effect of diffusion, as in Secs. 9 and 10.

APPENDIX B: THE ADDITION OF FLOW TERMS

Consider a collection of independent particles subject to a force field $\vec{F}(\vec{r})$. In the single particle phase space they constitute a cloud with density $n(\vec{r},\vec{v})$ obeying the equation

$$\frac{\partial n}{\partial t} = - \vec{v} \cdot \frac{\partial n}{\partial \vec{r}} - \vec{F}(\vec{r}) \cdot \frac{\partial n}{\partial \vec{v}} \equiv L\, n . \tag{B.1}$$

Obviously the product $n(\vec{r}_1,\vec{v}_1)\, n(\vec{r}_2,\vec{v}_2)$ obeys

$$d_t \left\{ n(\vec{r}_1,\vec{v}_1)\, n(\vec{r}_2,\vec{v}_2) \right\} = (L_1 + L_2) \left\{ n(\vec{r}_1,\vec{v}_1)\, n(\vec{r}_2,\vec{v}_2) \right\} , \tag{B.2}$$

where L_1 and L_2 are the same differential operator as in (B.1) but acting on $\vec{r}_1, \vec{v}_1$ and $\vec{r}_2, \vec{v}_2$ respectively.

Next consider an ensemble of such collections of particles. Obviously the ensemble average $\langle n(\vec{r},\vec{v}) \rangle$ of the particle density obeys the same equation (B.1) and $\langle n(\vec{r}_1,\vec{v}_1)\, n(\vec{r}_2,\vec{v}_2) \rangle$ obeys (B.2). It is easy to derive from this that the cumulant $\langle\!\langle n(\vec{r}_1,\vec{v}_1)\, n(\vec{r}_2,\vec{v}_2) \rangle\!\rangle$ also obeys (B.2).

Subsequently one obtains for the factorial cumulant

$$\begin{aligned} d_t[n(\vec{r}_1,\vec{v}_1)\, n(\vec{r}_2,\vec{v}_2)] &= \\ &= d_t \langle\!\langle n(\vec{r}_1,\vec{v}_1) n(\vec{r}_2,\vec{v}_2) \rangle\!\rangle - \delta(\vec{r}_1 - \vec{r}_2)\delta(\vec{v}_1 - \vec{v}_2)\, d_t \langle n(\vec{r}_1,\vec{v}_1) \rangle \\ &= (L_1 + L_2) \left\{ [n(\vec{r}_1,\vec{v}_1) n(\vec{r}_2,\vec{v}_2)] + \delta(\vec{r}_1 - \vec{r}_2)\delta(\vec{v}_1 - \vec{v}_2) \langle n(\vec{r}_1,\vec{v}_1) \rangle \right\} \\ &\quad - \delta(\vec{r}_1 - \vec{r}_2)\, \delta(\vec{v}_1 - \vec{v}_2)\, L_1 \langle n(\vec{r}_1,\vec{v}_1) \rangle . \\ &= (L_1 + L_2)\, [n(\vec{r}_1,\vec{v}_1)\, n(\vec{r}_2,\vec{v}_2)] + R . \end{aligned}$$

The remainder R is

$$(L_1+L_2)\ \delta(\vec{r}_1-\vec{r}_2)\delta(\vec{v}_1-\vec{v}_2)\langle n(\vec{r}_1,\vec{v}_1)\rangle - \delta(\vec{r}_1-\vec{r}_2)\delta(\vec{v}_1-\vec{v}_2)\ L_1 \langle n(\vec{r}_1,\vec{v}_1)\rangle$$

$$= \langle n(\vec{r}_1,\vec{v}_1)\rangle\ (L_1+L_2)\ \delta(\vec{r}_1-\vec{r}_2)\ \delta(\vec{v}_1-\vec{v}_2)\ .$$

However, L_1+L_2 acting on the delta functions equals

$$\left(\vec{v}_1\cdot\frac{\partial}{\partial\vec{r}_1}-\vec{v}_2\cdot\frac{\partial}{\partial\vec{r}_1}+\vec{F}(\vec{r}_1)\cdot\frac{\partial}{\partial\vec{v}_1}-\vec{F}(\vec{r}_2)\cdot\frac{\partial}{\partial\vec{v}_1}\right)\delta(\vec{r}_1-\vec{r}_2)\delta(\vec{v}_1-\vec{v}_2)=0.$$

Thus, in the case of pure flow it makes no difference whether one adds the flow terms to the second moment, the covariance, or the factorial cumulant. In the case of diffusion it is shown in Sec. 5 that it does make a difference, and it is necessary to add them to the factorial cumulant, in order to account for the stochastic character of the diffusion process.

APPENDIX C: CONNECTION WITH THE LANGEVIN EQUATION

Let u be a vector in a finite or infinite dimensional Banach space. Suppose it varies with time according to the linear equation

$$d_t u = A u + \varpi(t)\ . \tag{C.1}$$

A is a linear operator, which may depend on time. $\varpi(t)$ is a stochastic vector with the statistical properties

$$\langle\varpi(t)\rangle = 0\ , \qquad \langle\varpi(t_1)\tilde{\varpi}(t_2)\rangle = \delta(t_1-t_2)\ \Gamma(t_1), \tag{C.2}$$

where $\tilde{\varpi}$ is the transposed vector and Γ a symmetric matrix.

The Langevin equation (C.1) can be solved in terms of the evolution operator (or "propagator" or "Green's function") $Y(t|t')$ defined by

$$d_t\ Y(t|t') = A(t)\ Y(t|t'), \qquad Y(t|t) = 1\ . \tag{C.3}$$

The solution with given initial value u(0) is

$$u(t) = Y(t|0)\ u(0) + \int_0^t Y(t|t')\ \varpi(t')\ dt'\ .$$

It follows that the covariance matrix of u(t) is

$$\langle\!\langle u(t)\ \tilde{u}(t)\rangle\!\rangle = \int_0^t dt'\int_0^t dt''\quad Y(t|t')\langle\varpi(t')\varpi(t'')\rangle\ \tilde{Y}(t|t'')$$

$$= \int_0^t dt'\quad Y(t|t')\ \Gamma(t')\ \tilde{Y}(t|t')\ .$$

From this expression one derives a differential equation for the covariance matrix by inspection:

$$d_t \ll u(t)\,\tilde{u}(t) \gg - A \ll u(t)\,\tilde{u}(t) \gg - \ll u(t)\,\tilde{u}(t) \gg \tilde{A} = \Gamma(t)\,, \qquad \text{(C.4)}$$

where use has been made of (C.3). It is easily seen that the equation remains true when the covariance matrix $\ll u(t)\,\tilde{u}(t) \gg$ is replaced with the second moments matrix $\langle u(t)\,\tilde{u}(t)\rangle$.

Our applications of the master equation also led in each case to an equation of type (C.4) for the covariance matrix. We now see that the same result can be reproduced by a Langevin equation (C.1) with (C.2), provided Γ is chosen correctly.

For example, in Sec. 5 it was shown that the density of a diffusing substance obeys (5.11). It now follows that the same result can be obtained by supplementing the diffusion equation with a Langevin term

$$d_t n = D\nabla^2 n + \varpi(\vec{r},t)\,,$$

where $\langle \varpi(\vec{r},t)\rangle = 0$ and

$$\langle \varpi(\vec{r}_1,t_1)\,\varpi(\vec{r}_2,t_2)\rangle = 2D\;\delta(t_1-t_2)\;\nabla_1\cdot\nabla_2\;\;\delta(\vec{r}_1-\vec{r}_2)\;\langle n(\vec{r}_1)\rangle_t\,. \qquad \text{(C.5)}$$

Thus we have justified the use of a Langevin equation and derived the correlation function for the Langevin force by showing that it leads to the correct result, which has been derived in Sec. 5 using no other assumptions than those by which the macroscopic diffusion equation is derived.

In summary the conclusion from the calculation in this Appendix is that for linear or linearized equations every result obtained for the covariance can be reproduced by a Langevin equation (C.1), combined with a suitable choice for the matrix Γ. However, as appears in (C.5) the correct Γ may have to depend on time, and on the particular macroscopic solution for which one is investigating the fluctuations. It is therefore no easy matter to know a priori the correct form of Γ; except in equilibrium where the fluctuation-dissipation theorem can be utilized. This is why authors who use the Langevin approach for computing fluctuations in other than equilibrium states have to resort to an ad hoc and often rather doubtful guess about the form of Γ.

REFERENCES

1. N.G. van Kampen, Can. J. Phys. 39, 551 (1961); J. Math. Phys. 4, 190 (1963).
2. N.G. van Kampen, in: Fluctuation Phenomena in Solids (R.E. Burgess ed., Academic Press, New York 1965); also in: Adv. Chem. Phys. 15 (Interscience, New York 1969).
3. D.R. Mc Neil, Biometrika 59, 494 (1972); N.G. van Kampen, Biometrika 60, 419 (1973).
4. R. Kubo, K. Matsuo, and K. Kitahara, J. Statist. Phys. 9, 51 (1973).
5. N.G. van Kampen, in: Adv. Chem. Phys. (1975), to be published.
6. See e.g., L.D. Landau and E.M. Lifshitz, Statistical Physics (Pergamon, London 1958) ch. 12.

7. H.B.G. Casimir, Revs. Mod. Phys. 17, 343 (1945); S.R. de Groot and P. Mazur, Phys. Rev. 94, 218 and 224 (1954).
8. A. van der Ziel, Noise (Prentice-Hall, Englewood Cliffs, N.J. 1954).
9. A. van der Ziel, Fluctuation Phenomena in Semi-Conductors (Butterworths, London 1959); A. Gisolf and R.J.J. Zijlstra, Solid-State Electronics 16, 571 (1973).
10. K.M. van Vliet, A. Friedmann, R.J.J. Zijlstra, A. Gisolf, and A. van der Ziel, J. Appl. Phys. 46, 1804 and 1814 (1975).
11. P.J. Price, in: Fluctuation Phenomena in Solids (R.E. Burgess ed., Academic Press, New York 1965).
12. S.V. Gantsevich, V.L. Gurevich, and R. Katilyus, Soviet Physics – Solid State 11, 247 (1969).
13. W.A. Schlupp, Physik Kond. Materie 8, 167 (1968); 13, 307 (1971).
14. K.M. van Vliet and J.R. Fassett, in: Fluctuation Phenomena in Solids (R.E. Burgess ed., Academic Press, New York 1965).
15. V.A. Lo Dato, in: Probabilistic Methods in Applied Mathematics 3 (A.T. Barucha-Reid ed., Academic Press, New York 1973).
16. L.D. Landau and E.M. Lifshitz, Fluid Mechanics (Pergamon, Oxford 1959) ch. 17; R. Zwanzig, in: Proceedings of the 6th IUPAP Conference on Statistical Mechanics (S.A. Rice et al. eds., University of Chicago Press, Chicago 1972).
17. L.D. Landau and E.M. Lifshitz, Electrodynamics in Continuous Media (Pergamon, Oxford 1960) ch. 13; D. Langbein, in: Springer Tracts in Modern Physics 72 (1974).
18. H. Haken, Revs. Mod. Phys. 47, 67 (1975).
19. G. Nicolis and I. Prigogine, Proc. Nat. Acad. Sci. USA 68, 2102 (1971); G. Nicolis, J. Statist. Phys. 6, 195 (1972); G. Nicolis, P. Allen, and A. van Nypelseer, Prog. Theor. Phys. 52, 1481 (1974).
20. Y. Kuramoto, Prog. Theor. Phys. 49, 1782 (1973).
21. A. Nitzan and J. Ross, J. Statist. Phys. 10, 379 (1974).
22. Y. Kuramoto, Prog. Theor. Phys. 52, 711 (1974).
23. W. Weidlich, in: Cooperative Effects (H. Haken ed., North-Holland, Amsterdam 1974).
24. A. van der Ziel, Solid State Electronics 9, 123, 899, and 1139 (1966).
25. K.M. van Vliet, J. Mathem. Phys. 12, 1981 and 1998 (1971).
26. A. van der Ziel, Proc. IRE 43, 1639 (1955).
27. M.A. Lampert and P. Mark, Current Injection in Solids (Academic Press, New York 1970).
28. S.V. Gantsevich, V.L. Gurevich, and R. Katilius, Phys. Cond. Matter 18, 165 (1974).
29. A.J. Lotka, Elements of Mathematical Biology (Dover, New York 1956); E.W. Montroll, in: Quantum Theory and Statistical Physics (Boulder Lectures in Theoretical Physics 10A, A.O. Barut and W.E. Brittin eds., Gordon and Breach, New York 1968); E. Batschelet, Introduction to Mathematics for Life Scientists (Springer, Berlin 1971).

30. M.G. Kendall and A. Stuart, The Advanced Theory of Statistics, Volume I (Third edition, Griffin and Co, London 1969) p. 75.
31. H. Mori, Prog. Theor. Phys. 52, 433 (1974).
32. M. Malek-Mansour and G. Nicolis, J. Statist. Phys. 13, 197 (1975).
33. C.W. Gardiner, K.J. McNeil and D.F. Walls, Phys. Letters 53A, 205 (1975).
34. C.W. Gardiner, K.J. McNeil, D.F. Walls and I.S. Matheson, preprint.

Dr. van Kampen is Professor of Theoretical Physics at the University of Utrecht. He is a member of the Royal Netherlands Academy of Arts and Sciences. Some of his recent work has concerned stochastic differential equations and the theory of fluctuations in physical and biological systems. He first met Julius during a stay at the National Bureau of Standards and later, as Visiting Professor, worked with him at Howard University.

INTERACTIONS OF LIMIT CYCLE OSCILLATORS

Robert Zwanzig *
University of Maryland, College Park Md. 20742

ABSTRACT

Phenomena arising from the interaction of non-linear limit cycle oscillators are discussed. The individual oscillators are piecewise linear and discontinuous (the "playground swing" model). A special interaction is chosen, which allows much of the analysis to be performed algebraically rather than by solving differential equations. The model is well adapted to efficient numerical computation. Examples are discussed for which various types of synchronization are found.

INTRODUCTION

Limit cycle oscillations appear in the mathematical analysis of many phenomena. Among these are the theory of the laser, biochemical oscillators, circadian rhythms, and many engineering applications. Oscillators with two degrees of freedom (position x and velocity dx/dt) have been studied extensively; reference [1] provides a comprehensive survey. Systems of interacting limit cycle oscillators have been studied less extensively; in reference [2], Pavlides surveyed some of the literature. The qualitative methods that are available for systems of two degrees of freedom are no longer so useful, and numerical methods are often the only way to obtain useful information. We propose here a model system of interacting limit cycle oscillators for which much of the analysis can be performed algebraically, and for which numerical methods are expected to be particularly efficient.

First we discuss the individual oscillators, and show how easy it is to find useful information algebraically. Then we introduce a special interaction, chosen to maintain algebraic simplicity. We show that, in special cases, the motion of the oscillators can become perfectly synchronized as a result of their interaction. In other cases, various types of partial synchronization are found.

THE INDIVIDUAL OSCILLATOR

In order to focus attention on the phenomena induced by interactions, it is useful to start with individual oscillators which are as simple as possible, but which nevertheless show normal limit cycle behavior. We choose the "playground swing" model. The swing oscillates back and forth with its natural frequency, with normal

* Research supported by the National Science Foundation, under grant number MPS 73-08856 A02.

linear friction. At a certain point on its trajectory (for example the vertical position), a fixed impulse is given to the swing. Eventually, the swing settles down into a limit cycle oscillation; the amplitude is determined by the friction and by the given impulse. In reference (1) this is referred to as a "clock".

The equations of motion for this oscillator can be written in several forms. We start with two first order differential equations for the position y and the velocity v,

$$dy/dt = v \quad , \qquad\qquad (1)$$
$$dv/dt = -(1+f^2)y - 2fv + Q(y,v)$$

where $(1+f^2)^{1/2}$ is the natural frequency of the swing and $2f$ is the friction coefficient of damping. The quantity Q, which we call the impulse function, has the property that whenever y passes through 0, and v is positive, the velocity is suddenly increased by an amount A. (Clearly, the differential equations are autonomous.) In mathematical form, Q is

$$Q = A\ v\ S(v)\ \delta(y) \qquad (2)$$

where $S(v)$ is a step function, zero for negative v and unity for positive v, and $\delta(y)$ is the delta function. It is not hard to show that the time integral of Q through an impulse is just A.

For a more symmetrical form, we change variables,

$$x = v + fy \ . \qquad (3)$$

Then the equations of motion become

$$dx/dt = -y - fx + Q \quad , \qquad\qquad (4)$$
$$dy/dt = x - fy \quad ;$$

and Q becomes

$$Q = A\ x\ S(x)\ \delta(y) \ . \qquad (5)$$

Finally, we introduce the complex variable

$$z = x + iy \quad , \qquad (6)$$

so that

$$dz/dt = (i-f)z + Q \ . \qquad (7)$$

The signal for an impulse A is passage of the phase of z, or $\arg(z)$, through 2π. In between impulses, the motion is

$$z(t) = z(t') \exp(i-f)(t-t') \ . \qquad (8)$$

To find the limit cycle, we suppose that an impulse has just been delivered at time $t = 0$. (For convenience, we say that the oscillator has "fired".) Just after the impulse, the state of the oscillator is $z(0)$. It moves freely, according to Eq. (8), until the next firing, which comes at $t = 2\pi$. The next impulse increases z by an amount A. For a limit cycle, the new z must be the same as the old one, or

$$z(0) = A + z(0) \exp(-2\pi f) . \tag{9}$$

This equation gives the amplitude $z(0)$ of the limit cycle in terms of A and f. In the following, we choose the impulse

$$A = 1 - \exp(-2\pi f) \tag{10}$$

so that the amplitude is unity. The shape of the limit cycle is sketched in Fig. (1).

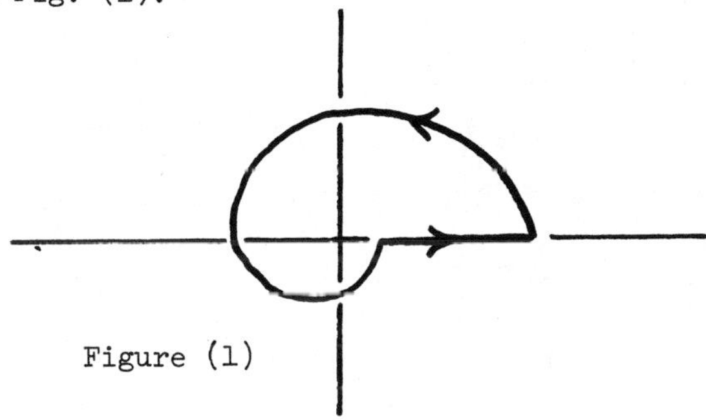

Figure (1)

If z is close to unity at $t = 0$, then after one complete cycle it will be even closer; this can be seen by a simple linearization of the relation between z's at successive firings. Thus the limit cycle is stable.

The response of the oscillator to an external disturbance, and the phenomenon of entrainment, are often conveniently discussed using the phase response curve of the oscillator. This technique is due particularly to Pittendrigh and to Winfree; extensive references are given by Pavlides. Without going into the details, we observe that the phase response curve of the "playground swing" oscillator can be found easily by algebraic means. This is another example of the great simplicity of the model.

A MODEL FOR INTERACTING OSCILLATORS

The simplicity of the "playground swing" model is due to three features. (1) Between impulses, its motion is described by a simple linear differential equation. (2) The timing of the impulses is determined autonomously by a signal, passage of the phase of z through a specified value. (3) The impulse itself is independent of the state of the oscillator. We want to construct a model for interacting oscillators which preserves these features.

First, we write the equations of motion for the interacting oscillators, labelled by $j = 1, 2, \ldots N$ in the form

$$dz_j/dt = (i-f)z_j + Q_j \tag{11}$$

where Q_j is the impulse function for the j-th oscillator. Next, we introduce a signal function W_j which is a function of the z's,

$$W_j = U_j + iV_j \quad . \tag{12}$$

In analogy with Eq. (5), we define the impulse function by

$$Q_j = A\, U_j\, S(U_j)\ \delta(V_j) \ , \tag{13}$$

so that the signal for firing of the j-th oscillator comes when the phase of W_j passes through 2π .

In the special case

$$W_j = z_j \tag{14}$$

we have a system of N non-interacting oscillators. Other choices for W will describe interacting oscillators. In particular, the choice

$$W_j = \sum a_{jk} z_k \tag{15}$$

seems appealingly simple, and we shall restrict our attention to this one.

The motion of this system can be found by the following sequence of operations. Suppose that some oscillator has fired at time s ; just after this, the values of the z's are given by $z_j(s)$. Until the next impulse, their time dependence is known,

$$z_j(t) = z_j(s) \exp(i-f)(t-s) \quad . \tag{16}$$

Then the time dependence of the signal function is also known,

$$W_j(t) = \sum a_{jk}\, z_k(t) \ . \tag{17}$$

We scan the signal functions for $j = 1, 2, \ldots N$, and find the one with the largest phase (modulo 2π); this one is $j = J$.

Then the J-th oscillator will be the next one to fire, and the phase of W_J determines when this will occur. Just after firing, z_J is replaced by z_J+A, and the other z's are unchanged. Now the above sequence of operations is repeated to determine the identity of the next oscillator to fire, and the time at which that occurs.

Evidently, this sequence is ideally adapted to numerical computation. It is not necessary to solve differential equations numerically; only algebraic operations are involved. The model has a close resemblance to the hard sphere model of molecular dynamics. In the hard sphere model, molecules move in straight lines between collisions. One has to scan the relative motions of pairs of molecules to decide which pair will collide next; this is algebraic. Then one has to find the velocities of that pair after the collision; this also is algebraic. As in the hard sphere system, it would probably be reasonable to handle several hundred interacting oscillators numerically.

SIMPLE EXAMPLE OF SYNCHRONIZATION

As a particularly simple example of oscillator synchronization, we discuss first the signal function

$$W_j = \sum z_j = W . \tag{18}$$

All oscillators have an equal effect on each other. Because W_j does not depend on the index j, the same is true of the impulse functions,

$$Q_j = Q . \tag{19}$$

On subtracting the equations of motion of the j-th and k-th oscillators, the impulse functions cancel:

$$d(z_j-z_k)/dt = (i-f)(z_j-z_k) . \tag{20}$$

All differences decay to zero at large t, and all oscillators come into perfect synchronization. In fact, they all move simultaneously along the limit cycle of the individual oscillator, with period 2π.

SYMMETRIC INTERACTION OF TWO OSCILLATORS

Not quite as simple as the preceding example, but still simple enough to allow a fairly detailed analysis, is the case of two oscillators with symmetric coupling. The signal functions are

$$\begin{aligned} W_1 &= az_1 + bz_2 , \\ W_2 &= bz_1 + az_2 , \end{aligned} \tag{21}$$

where a and b are arbitrary real numbers. It turns out that when these are both positive, there is a locally stable limit cycle of period 2π in which the oscillators are synchronized. If a is positive and b is negative, there is a locally stable limit cycle of period 2π in which the oscillators are anti-synchronized. If a is negative, there are also unstable limit cycles.

To illustrate how this system can be handled algebraically, we investigate solutions of period 2π in which the oscillators fire alternately. The procedure is to follow the sequence of signal functions, to investigate the possibility of periodic solutions, and then to check on the stability of such solutions.

Suppose that just before $t = 0$, oscillator #1 is about to fire. Then the signal functions must have the form

$$W_1 = U_1 - i0 \ ; \quad U_1 > 0 \ ; \qquad W_2 = \text{arbitrary}. \tag{22}$$

Just after firing, the oscillator states are changed,

$$z_1(+0) = z_1(-0) + A \ , \qquad z_2(+0) = z_2(-0) \ . \tag{23}$$

The signal functions also change,

$$W_1(+0) = U_1 + aA = \omega_1 \ , \qquad W_2(+0) = W_2 + bA \ . \tag{24}$$

Presumably the signal function W_2 now has a larger phase than W_1, so that oscillator #2 will fire next. Let this happen at t'. Just before t', the signal functions are

$$W_1(t'-0) = G(t') \ \omega_1 \ , \qquad W_2(t'-0) = G(t')(W_2 + bA) \ , \tag{25}$$

where we abbreviate

$$G(t) = \exp(i-f)t \ . \tag{26}$$

At $t'-0$, W_2 has a real positive part and an infinitesimal negative imaginary part; this condition will be used shortly. After #2 fires, the states are

$$z_1(t'+0) = z_1(t'-0) \ , \qquad z_2(t'+0) = z_2(t'-0) + A \ . \tag{27}$$

The corresponding signal functions are

$$W_1(t'+0) = G(t')\ \omega_1 + bA\ , \qquad W_2(t'+0) = G(t')(W_2 + bA) + aA = \omega_2\ . \tag{28}$$

Now #1 will fire next, at time t''. The signal functions before and after this are

$$W_1(t''-0) = G(t''-t')(G(t')\ \omega_1 + bA)\ , \qquad W_2(t''-0) = G(t''-t')\ \omega_2\ , \tag{29}$$

and

$$W_1(t''+0) = aA + G(t''-t')(G(t')\ \omega_1 + bA)\ , \qquad W_2(t''+0) = bA + G(t''-t')\ \omega_2\ . \tag{30}$$

Having worked out a complete sequence, we now require periodicity:

$$W_1(t''+0) = W_1(+0)\ , \qquad W_2(t''+0) = W_2(+0)\ . \tag{31}$$

After some algebraic manipulation, this is equivalent to the conditions

$$\omega_1 - aA = G(t''-t')bA + G(t'')\omega_1\ , \qquad \omega_2 - aA = G(t')bA + G(t'')\omega_2\ . \tag{32}$$

The condition that the oscillators fire alternately requires that the left hand sides of Eq. (32) are real and positive. We have here two complex equations for the four unknown quantities ω_1, ω_2, t', t''.

The following procedure provides solutions of Eq. (32). We multiply the first equation by exp(-it'') , and then take the real and imaginary parts of both equations. We denote cos t' by C' , sin t' by S' , exp(-ft') by q' , and similarly for the doubly primed quantities. The four equations become

$$bA(q''/q')C' - (\omega_1 - aA)C'' = -\omega_1 q''$$
$$bAq'C' + \omega_2 q''C'' = \omega_2 - aA$$
$$bA(q''/q')S' - (\omega_1 - aA)S'' = 0 \tag{33}$$
$$bAq'S' + \omega_2 q''S'' = 0\ .$$

If the determinant of the last two equations does not vanish, then we may conclude that

$$S' = S'' = 0 , \tag{34}$$

so that t' and t'' are restricted to multiples of π . This condition on the determinant can be verified at the end of the calculation. The first two equations determine values for ω_1 and ω_2. The condition that the left hand sides of Eq. (32) must be real and positive places limits on the allowed values of a and b.

For example, take t' = t'' = 2π . We denote $\exp(-\pi f)$ by q (without any primes). Then

$$\begin{aligned} \omega_1 &= a+b , \\ \omega_2 &= a+bq^2 . \end{aligned} \tag{35}$$

(Recall that $A = 1-q^2$.) Then a and b are limited by

$$\begin{aligned} a+b &> 0 , \\ b+aq^2 &> 0 . \end{aligned} \tag{36}$$

If a is positive, these conditions are met for all positive b and for some negative b . The resulting motion, if it is stable, corresponds to perfect synchronization of the oscillators, since they both fire at the same time.

The choice t' = π and t'' = 2π leads to

$$\omega_1 = \omega_2 = a-bq \tag{37}$$

and to the condition

$$b < aq . \tag{38}$$

If a is positive, this condition is met for all negative b and for some positive b . If stable, it corresponds to anti-synchronization since they fire with the phase difference π .

The choice t' = t'' = π does not lead to an acceptable solution for positive a .

We see that there is a region of the (a,b) plane in which either synchronization or anti-synchronization can occur. Which will one actually find? This can be decided by an analysis of stability.

To Illustrate this analysis, we consider the first case, where t' = t'' = 2π . Further, we set a = 1 (only the ratio b/a is significant). The procedure is to study the sequence of signal functions, as in Eqs. (22-30), starting with a state near the limit cycle.

On the limit cycle, $W_1 = b+q^2$ and $W_2 = 1+bq^2$. So we start with

$$W_1(-0) = b+q^2 + \varepsilon_1 - i0 \ , \qquad (39)$$
$$W_2(-0) = 1+bq^2 + \varepsilon_2 + i\,\delta_2 \ ,$$

where ε_1, ε_2, and δ_2 are sufficiently small that we can linearize all expressions with respect to them. Then we follow the sequence of firings of #1 and #2, ending at t''-0. The resulting signal functions have the same general form as the initial ones, but with new values for ε and δ . These are related to the old ones by

$$(\varepsilon_1)'' = q^2 \varepsilon_1 \ ,$$
$$(\varepsilon_2)'' = q^2\varepsilon_2 + q^2 f\,\delta_2 - q^2 f(1+bq^2)/(b+q^2)\,\delta_2 \ , \qquad (40)$$
$$(\delta_2)'' = q^2(1+bq^2)/(b+q^2)\,\delta_2 \ .$$

Since q is smaller than unity, ε_1 tends to zero. According to the third condition, δ_2 tends to zero if b is positive, but increases if b is negative. If b is positive, then ε_2 also tends to zero. This shows that the synchronized solution is locally stable if both a and b are positive.

The same kind of analysis shows that the anti-synchronized solution is locally stable if a is positive and b is negative. (In the special case b = 0 , the oscillators are uncoupled and each maintains its own limit cycle; the phase difference is arbitrary.) If a is negative and b is positive, there are regions of the (a,b) plane in which one finds synchronized solutions, and regions of unstable behavior. If a and b are both negative, one finds anti-synchronized or unstable behavior.

WEAKLY NONLINEAR SYSTEMS

Further progress can be made in the analysis of systems of many interacting oscillators when they are weakly nonlinear. By this we mean that the impulse A is very small; with our earlier choice $A = 1 - \exp(-2\pi f)$, this is equivalent to taking very small f .

Starting with Eq. (11), we substitute

$$z_j = \exp(it)\, g_j \qquad (41)$$

so that

$$dg_j/dt = -fg_j + \exp(-it)\, Q_j \ . \qquad (42)$$

We write the signal function in the form

$$W_j = e^{it}\, r_j\, e^{i\phi_j t} \tag{43}$$

which introduces an amplitude r_j and a phase ϕ_j, both functions of time. By using properties of the delta and step functions, it is easy to see that the impulse function becomes

$$Q_j = A \sum \delta(t+\phi_j - 2\pi n) \tag{44}$$

where n is any integer. The impulse function vanishes except when the phase of the signal function passes through an integral multiple of 2π, and its integral over an impulse is A.

In the weakly nonlinear limit, A is approximately $2\pi f$, so that the rate of change of g_j is of the order of f and is small. This suggests use of the Kryloff-Bogolyubov procedure. Thus we average Eq. (42) over one cycle, of length 2π. (From here on we do not distinguish between g_j and its average-- it must be remembered that all equations are concerned only with the average.) The resulting averaged equation is

$$dg_j/dt = -fg_j + f\, e^{i\phi_j} \quad . \tag{45}$$

On the limit cycle, g_j is constant. Thus the condition for limit cycles is

$$g_j = e^{i\phi_j} \quad . \tag{46}$$

Solutions of these equations will be labelled by the symbol $\hat{}$, as in $\hat{g}_j$, $\hat{\phi}_j$.

Now let us return to the signal function W_j:

$$W_j = \sum a_{jk}\, e^{it}\, g_k = e^{it}\, r_j\, e^{i\phi_j} \quad . \tag{47}$$

By using Eq. (46), we see that the conditions for a limit cycle are

$$\hat{r}_j\, e^{i\hat{\phi}_j} = \sum a_{jk}\, e^{i\hat{\phi}_k} \quad . \tag{48}$$

This is supplemented by the condition that all $\hat{r}_j$ are positive.

What can be said about the stability of such limit cycles? Let us expand in small deviations h_j from the limit cycle,

$$g_j = e^{i\hat{\phi}_j}\, (1+h_j) \tag{49}$$

and then linearize in the deviations. The result is

$$f^{-1}\, dh_j/dt = -h_j + i\, \mathrm{Im}\, [\, \sum (a_{jk}/\hat{r}_j)\, e^{i(\hat{\phi}_k-\hat{\phi}_j)}\, h_k\,] \quad . \tag{50}$$

Clearly, the real parts of all h_j decay exponentially to zero; they are not coupled to the imaginary parts. We define

$$\varepsilon_j = \text{Im } h_j \qquad (51)$$

so that

$$f^{-1} d\varepsilon_j/dt = -\varepsilon_j + \sum (a_{jk}/\hat{r}_j) \cos(\hat{\phi}_k - \hat{\phi}_j) \; \varepsilon_k \,. \qquad (52)$$

Then the stability matrix $\underset{\sim}{S}$ is given by

$$S_{jk} = (a_{jk}/\hat{r}_j) \cos(\hat{\phi}_k - \hat{\phi}_j) - \delta_{jk} \quad ; \qquad (53)$$

for stable (semi-stable) limit cycles, the eigenvalues of $\underset{\sim}{S}$ must have negative (non-positive) real parts.

To illustrate the above analysis, we consider an infinite one-dimensional lattice of interacting limit cycle oscillators, with nearest neighbor coupling. The interaction matrix is

$$a_{jk} = a\,\delta_{jk} + b(\delta_{j,k+1} + \delta_{j,k-1}) \,. \qquad (54)$$

The condition for limit cycles, Eq. (48), has the solution

$$\phi_j = \phi_0 + \theta j \qquad (55)$$

where ϕ_0 and θ are arbitrary angles. The amplitudes r_j are

$$r = r_j = a + 2b \cos\theta \quad (\text{all } j). \qquad (56)$$

We have a limit cycle for any θ for which r is positive. For given θ, the stability matrix has the eigenvalues (labelled by q) :

$$\lambda(\theta;q) = -[2b/r(\theta)] \cos \qquad (1-\cos q) \,. \qquad (57)$$

If the θ - cycle is stable, these must be negative for all q. Thus, conditions for existence and stability are

$$\begin{aligned} a + 2b\cos\theta &> 0 \,, \\ b\cos\theta &> 0 \,. \end{aligned} \qquad (58)$$

In particular, if a and b are both positive, then limit cycles with

$$|\theta| < \pi/2 \qquad (59)$$

are locally stable, and all others are unstable. Stable limit cycles are characterized by phases [Eq. (55)] which vary uniformly along the lattice. It would be of some interest to investigate the effects of noise in causing transitions between the various locally stable limit cycles, but this is a job for the future.

The same kind of analysis can be carried out for more complex networks. For example, suppose that

$$a_{jk} \geq 0 \tag{60}$$

for all j and k , and that there is at least one non-vanishing element in each row. Then the fully synchronized solution

$$\hat{\phi}_j = \hat{\phi} \quad (\text{all } j) \tag{61}$$

is locally stable. Equation (48) gives

$$\hat{r}_j = \sum a_{jk} > 0 \ . \tag{62}$$

The stability matrix for this limit cycle is

$$S_{jk} = (a_{jk}/\hat{r}_j) - \delta_{jk} \quad . \tag{63}$$

But $a_{jk}/\hat{r}_j$ is a row-stochastic matrix and the real parts of its eigenvalues are all less than or equal to unity. Thus the eigenvalues of $\underset{\sim}{S}$ satisfy conditions for semi-stability. A more careful investigation indicates that this solution is in fact locally stable. However, as we saw in the preceding example, there may be other locally stable limit cycles as well.

From the above discussions, it should now be clear that this model for interacting "playground swing" oscillators has great potential value for the analysis of complex interaction schemes.

REFERENCES

1. A. A. Andronov, A. A. Vitt, and S. E. Khaikin, Theory of Oscillators, Pergamon Press Ltd., 1966
2. T. Pavlidis, Biological Oscillators: Their Mathematical Analysis, Academic Press, New York, 1973

Dr. Zwanzig is Research Professor at the Institute for Fluid Dynamics and Applied Mathematics and at the Institute for Molecular Physics at the University of Maryland. He was Senior Theoretical Physical Chemist at the National Bureau of Standards and is a member of the National Academy of Sciences. His recent work has concerned the theory of both equilibrium and non-equilibrium processes in liquids and gases. He met Julius at the Office of Naval Research and they were associated for many years at the National Bureau of Standards.

Appendix

A. List of publications of Julius L. Jackson

B. The Car Parking Problem: A Reminiscence of Julius Jackson; F. Gornick

Publications

1. M.E. Rose and J.L. Jackson, "Ratio of L_1 to K Capture," Phys. Rev. 76, 1540 (1949).

2. J.L. Jackson, "A Variational Approach to Nuclear Reactions," Phys. Rev. 83, 301 (1951). (Taken in part from PH.D. Thesis; N.Y.U., 1950).

3. J.L. Jackson, "A Note on 'Irreversibility and Generalized Noise'," Phys. Rev. 87, 471 (1952).

4. H.B. Callen, M.L. Barasch, and J.L. Jackson, "Statistical Mechanics of Irreversibility," Phys. Rev. 88, 1382 (1952).

5. J.L. Jackson and M.C. Yovits, "Further Properties of the Quantum Statistical Impedance," Phys. Rev. 96, 15 (1954).

6. M.C. Yovits and J.L. Jackson, "Linear Filter Optimization with Game Theory Considerations," Bulletin of the IRE, Part 4, Computers, Information Theory, and Automatic Control, p. 193 (1955).

7. J.L. Jackson, "Stabilized Free Radicals," J. Wash. Academy of Sciences 48, 181 (1958).

8. J.L. Jackson and E.W. Montroll, "Free Radical Statistics," J. Chem. Phys. 28, 1101 (1958).

9. V. Griffing, J.L. Jackson and B.J. Ransil, "The Magnetic Interaction of H_3," J. Chem. Phys. 30, 1066 (1959).

10. J.L. Jackson, "Dynamic Stability of Frozen Radicals. I. Description and Application of the Model," J. Chem. Phys. 31, 154 (1959).

11. J.L. Jackson, Dynamic Stability of Frozen Radicals. II. The Formal Theory of the Model," J. Chem. Phys. 31, 722 (1959).

12. J.L. Jackson, "Reactions en Chaine de Radicaux Gélés," J. de Chimie Physique 56, 771 (1959).

13. J.L. Jackson, "Free Radical Trapping--Theoretical Aspects," Chapter 10 of Formation and Trapping of Free Radicals (edited by Bass and Broida, New York, Academic Press, 1960).

14. J.L. Jackson, "Electric Field Distribution in a Dense Plasma," Physics of Fluids, 3, 927 (1960).

15. S. Lifson and J.L. Jackson, "On the Self-Diffusion of Ions in a Polyelectrolyte Solution," J. Chem. Phys. 36, 2410 (1962).

16. J.L. Jackson, D.B. Levine and R.A. Piccirelli, "Plasma with Net Charge," Phys. Fluids 5, 250 (1962).

17. F. Gornick and J.L. Jackson, "The Sequence Selection Problem in the Crystallization of Polymers," J. Chem. Phys. 38, 1150 (1963).

18. J.L. Jackson and S.R. Coriell, "Effective Diffusion Constant in a Polyelectrolyte Solution," J. Chem. Phys. 38, 959 (1963).

19. S.R. Coriell and J.L. Jackson, "Calculations of the Potential and Effective Diffusion Constant in a Polyelectrolyte Solution," NBS Technical Note 192 (June 28, 1963).

20. S.R. Coriell and J.L. Jackson, "The Potential and Effective Diffusion Constant in a Polyelectrolyte Solution," J. Chem. Phys. 39, 2418 (1963).

21. J.L. Jackson and L. Klein, "Continuum Description of a High Temperature Plasma," Proc. VI International Conference on Ionization Processes in Gases (Paris, 1963), Vol. II, No. 11, 215-15 (1964).

22. J.L. Jackson and L. Klein, "Potential Distribution Method in Equilibrium Statistical Mechanics," Physics of Fluids 7, 228, (1964).

23. J.L. Jackson and L. Klein, "Continuum Theory of a Plasma," Physics of Fluids 7, 232 (1964).

24. J.L. Jackson and S.R. Coriell, "On Associated Ions in Polyelectrolytes and Trapped Brownian Trajectories," J. Chem. Phys. 40, 1490 (1964).

25. J.L. Jackson and S.R. Coriell, "On Trapped Trajectories in Brownian Motion," J. Math. Phys. 5, 1075 (1964).

26. J.L. Jackson and P. Mazur, "On the Statistical Mechanical Derivation of the Correlation Formula for the Viscosity," Physica 30, 2295 (1964).

27. P.G. Klemens and J.L. Jackson, "Diffusion of Electrons on the Fermi Surface. I. Lorenz Number of Monovalent Metals," Physica 30, 1780 (1964).

28. P.G. Klemens and J.L. Jackson, "Diffusion of Electrons on the Fermi Surface. II. Longitudinal Magnetoresistance of Ib Metals," Physica 31, 1421 (1965).

29. J.L. Jackson, M.C. Shen and D.A. McQuarrie, "Interchain Obstruction in Rubber Elasticity Theory," J. Chem. Phys. 44,

233 (1966).

30. M.C. Shen, D.A. McQuarrie and J.L. Jackson, "Thermoelastic Behavior of Natural Rubber," J. Appl. Phys. 38, 791 (1967).

31. S.R. Coriell and J.L. Jackson, "The Probability Distribution of the Radius of Gyration of an Ideal Polymer," J. Math. Phys. 8, 1276 (1967).

32. J.L. Jackson and L.S. Klein, "Statistical Mechanics of a Partially Ionized Plasma," IUPAP Int. Conf. on Statistical Mechanics, edited by Thor Bak, pp. 275-283, W.A. Benjamins, N.Y. 1967.

33. J.L. Jackson and L.S. Klein, "The Saha Equation in the Debye-Huckel Approximation," 8th Int. Conf. on Ionized Gases," p. 285, Vienna, Austria; Aug. 1967.

34. J.L. Jackson and S.R. Coriell, "Transport Properties of Composites," J. Appl. Phys. 39, 2349 (1968).

35. S.R. Coriell and J.L. Jackson, "Bounds on Transport Coefficients of Two-Phase Materials," J. Appl. Phys. 39, 4733 (1968).

36. J.L. Jackson and S.R. Coriell, "Continuum Description of an Equilibrium Van der Walls Gas," Int. Conf. on Statistical Mechanics, Kyoto, Japan, Sept. 9-14, 1968.

37. J.L. Jackson and L.S. Klein, "Equilibrium Theory of a Partially Ionized Plasma," Phys. Rev. 177, 352 (1969).

38. J.L. Jackson, S. Lifson and S.R. Coriell, "Association Time of Counterions to Polyelectrolytes in Solution," Comment, J. Chem. Phys., 50, 5045(L) (1969).

39. S.R. Coriell and J.L. Jackson, "The Probability Distribution of the Radius of Gyration for Two Statistical Segments," J. Chem. Phys. 53, 3389 (1970).

40. J.R. Creighton and J.L. Jackson, "Simplified Theory of Picosecond Pulses in Lasers," J. Appl. Phys. 42, 3409 (1971).

41. J.L. Jackson, "Further Considerations on the Theory of Inter molecular Obstruction in Rubber Elasticity," J. Chem. Phys. 12, 5124 (1972).

42. L.A. Segel and J.L. Jackson, "Dissipative Structure: An Explanation and an Ecological Example," Biology, J. Theor. Biol. 37, 545 (1972).

43. J.L. Jackson and A. Oplatka, "A Mechanochemical Theory of

Fluctuations in Muscles: I. Introduction to the Theory," Biorheology 11, 315 (1974).

44. J.L. Jackson, "Charge Neutrality in Electrolytic Solutions and the Liquid Junction Potential," J. Phys. Chem. 78, 2060 (1974).

45. J.L. Jackson and M.P. Shaw, "The Form and Stability of Current-Voltage Characteristics for Ideal Thermal Switching," Appl. Phys. Lett. 25, 666 (1974).

46. J.L. Jackson and M.P. Shaw, "An Ideal Model for Switching in Thin VO_2 Films," J. Solid State Chem. 12, 408 (1975).

Supplementary List

1. J.L. Jackson, "On the Continuity of the First Derivative in the Wigner-Eisenbud Formalism," Phys. Rev. 76, 185 (1949).

2. J.L. Jackson, "On Approximate Relativistic Wave Functions and Internal Pair Formation," Oak Ridge Report ORNL-746, Phys. General, August (1949).

3. J.L. Jackson, "Determining Yaw, Pitch, and Roll Angular Rates for Homing Missiles with Free Gyroscopes," APL/JHU Internal Memo (BDA), August (1951).

4. J.L. Jackson, "Interferometer Homing on a Line Reflector," JPL/JHU Internal Memo (BDA), January (1952).

5. J.L. Jackson, "Simplified Model for the Acquisition Problem," APL/JHU Internal Memo (BDA), June (1952).

6. J.L. Jackson, "Noise Due to Random False Code Pulses," APL/JHU CM-769, January (1953).

7. J.L. Jackson, "Game Theory Solution of the Filtering Problem," APL/JHU Internal Memo (BDA), July (1953).

8. J.L. Jackson, "Properties of the Rayleigh Distribution," APL/JHU CF-2155, January (1954).

9. J.L. Jackson and K. Froehlich, "Acquisition Computations," APL/JHU Internal Memo (BDA), May (1954).

10. J.L. Jackson, "Equations of Motion of Coupled Ordered and Dissipative Systems," Phys. Rev. 95, 642 (1954). Abstract.

11. J.L. Jackson, "Non-Minimum Phase Optimum Filters," APL/JHU Internal Memo (BDA), July (1954).

12. J.L. Jackson, "Spectrum Analysis of Boxcarred Functions,"

APL/JHU Internal Memo (BDA), December (1954).

13. J.L. Jackson, "Optimum Acquisition Procedure," APL/JHU Internal Memo (BDA), March (1955).

14. J.L. Jackson and K. Froehlich, "Optimum Acquisition Procedure," APL/JHU CM-861, January (1956).

15. E. Callen, H.B. Callen and J.L. Jackson, "Temperature Dependence of the Magnetocrystalline Anistropy Coefficients in Cubic Crystals," Bull. Amer. Phys. Soc., Sept. 1957 (Denver meeting). Also Conf. on Magnetism and Magnetic Materials, Washington, D.C., November (1957).

16. J.L. Jackson, "The Transient Behavior of a Dissipative System," NBS Report 5497, October (1957). Also, Bull. of Amer. Phys. Soc., January (1958), N.Y. meeting.

17. J.L. Jackson, "Stabilized Free Radicals," in Proceedings Advanced Propulsion Systems. (New York: Pergamon Press), pp. 205-11. (1959).

18. J.L. Jackson, "Calculations of $V(t)$ and $V^2(t)$ with Random Reflections and Sinusoidal Driving Forces," Convair San Diego Report, Microwave Energy Transfer Mechanisms; December (1960), pp. 99-106.

19. J.L. Jackson and S. Borowitz, "Greenhouse Effects in Radiation Absorption in Plasmas," Convair San Diego Report, Microwave Energy Transfer Mechanisms; December (1960), pp. 213-217.

20. J. Menkes and J.L. Jackson, "On the Propagation at a Weak Pressure Wave at Very Low Pressures," IDA Res. Paper, p. 328, September (1967).

21. J.L. Jackson and B. Skates, "Anisotropic Obstruction in Rubber Elasticity Theory--The Next Order," Bulletin of the APS, Sec. II, 13, 496 (1968).

22. J.L. Jackson, "A Note on the Geometrical Effect on the Hall Resistance," United Aircraft Research Lab Report UARG 42, March (1968).

23. J.L. Jackson, "Ideal Avalanche Noise," United Aircraft Research Lab Report UARG 61, April (1968).

24. J.L. Jackson, "The Spectrum of FM Noise," United Aircraft Research Lab Report UARG 75, June (1968).

25. J.L. Jackson, Chapter I, "Statistical Properties of Random Functions Utilized in Modeling Electron Density Fluctuations,"

of IDA Research Paper, p. 441, "Topics in Electromagnetic Scattering From a Random Plasma Slab," August (1968).

26. J.R. Creighton and J.L. Jackson, "A Simplified Theory of Picosecond Pulses in Lasers," Lawrence Radiation Laboratory, UCRL-71984, September 19, 1969.

27. J.L. Jackson, "Relativistic Treatment of a Corner-Reflected Signal," October (1969).

THE CAR PARKING PROBLEM: A REMENISCENCE OF JULIUS JACKSON

Fred Gornick
Chemistry Dept., University of Maryland Baltimore Co.

In the spring of 1960, when I was a postdoctoral fellow at the National Bureau of Standards in Washington, I encountered a combinatorial problem related to a then current model for polymer crystallization. My attempts to find a closed form solution to the problem had been fruitless and a colleague advised me to enlist the aid of the Statistical Physics Section of the Bureau, then headed by Melville Green. Included in this group were such accomplished theorists as Irwin Oppenheim, Robert Zwanzig and Julius Jackson. I suspect that my colleague's counsel to see Julius first was perhaps based on a feeling that he was less likely to eject me from the room than were the other tigers.

The combinatorial problem proved to be more difficult than I had thought and my discussion of it with Julius aroused his interest in it and led ultimately to a paper we published in 1962[(1)] in which an approximate solution was presented. More important to me than the paper was that our association led to a warm personal friendship which lasted until his death. Those who knew Julius could not help but be impressed by his immense capacity for friendship and for his ability to sustain close ties with people well beyond the circle of his immediate associates. Moreover, he was not content to merely have legions of friends; he also wanted his friends to be each other's friends. In consequence, I came to know most of the members of the Statistical Physics Section who, as it turned out, were not such tigers after all.

The actual physical background of the combinatorial problem on which Julius and I worked need not concern us here. It can be stated in simpler terms as follows: Cars of uniform length are permanently parked at random along a curb of infinite length. If two adjacent cars are parked such that the distance between them is less than the length of a single car, then that space is unavailable for further parking. Two questions are asked: (1) What is the distribution of intervals between parked cars as a function of the fraction of curb space utilized? (2) What is the limiting value of the expected fraction of wasted space when the curb is saturated? An approximate answer to the second question had been obtained by the Hungarian mathematician, A. Renyi,[(2)] who employed a rather complex numerical method leading to a value of 0.75 as the limiting fraction of occupied space. Julius and I were impressed by Renyi's solution which required that the problem be treated sequentially, the entire past history of the system being taken into account at each step, i.e. as each successive parked car selected an available space of sufficient size. We were interested in comparing Renyi's result with that obtained from a less realistic but more mathematically tractable model, namely one that allowed for re-parking or shifting of cars following their initial positioning. In the latter case no saturation value could be obtained except in a rather oblique manner. Denoting the fraction of curb space occupied by parked cars as θ and the fraction of space available for further parking as W, our model led to the expression

$$W = 1-\theta-\exp[-\theta/(1-\theta)]$$

The real parking problem requires that W approach zero as θ approaches its saturation value of approximately 0.75 whereas, according to our expression W is always greater than zero as long as θ is less than unity. It is however noteworthy that W attains its maximum value at $\theta=0.715$ and decreases monotonically thereafter. We interpreted this result, which differed from Renyi's by 5%, as an effective upper bound on θ, above which our approximation could not be expected to hold. A more physical interpretation of this result might be that the attainment of a maximum value of W corresponds to the point where the moving of previously parked cars becomes a necessary condition for further occupancy.

The car parking problem remained of interest to us for years after we published our paper on it. I recall one occasion in Julius' home when we tried to obtain some experimental results by marking off line segments on a piece of paper. It was a do-it-yourself Monte-Carlo method. About a year before his death Julius sent me a draft of a short paper he had written on a related problem. It dealt with a process in which point nuclei are formed at random along a line segment and proceed to expand until all available space is filled, the growth centers being confined only to unoccupied space. It was a variant on the car parking problem. I made some comments on the paper and a suggestion regarding relevant references. Julius was preparing for his last trip to Israel and the paper was set aside. It was never published.

1/ F. Gornick, J. L. Jackson, J. Chem. Phys. Vol. 38, 1150-54 (1963)

2/ A. Renyi, Commun. Math Research Inst. Hungarian Acad. Sci. Vol. 3, 129 (1958).

Dr. Gornick is Professor of Chemistry and former Chairman of the Department of Chemistry at the University of Maryland, Baltimore County. He was a member of the Materials Science Department of the University of Virginia and staff member of the National Bureau of Standards. His recent interests have been in the kinetics and mechanisms of phase transitions in macro-molecular systems. He first met Julius when they were colleagues at the National Bureau of Standards.

AIP Conference Proceedings

No.	Title	L. C. Number	ISBN
No. 1	Feedback and Dynamic Control of Plasmas (Princeton) 1970	70-141596	0-88318-100-2
No. 2	Particles and Fields - 1971 (Rochester)	71-184662	0-88318-101-0
No. 3	Thermal Expansion - 1971 (Corning)	72-76970	0-88318-102-9
No. 4	Superconductivity in d- and f-Band Metals (Rochester 1971)	74-18879	0-88318-103-7
No. 5	Magnetism and Magnetic Materials - 1971 (2 parts) (Chicago)	59-2468	0-88318-104-5
No. 6	Particle Physics (Irvine 1971)	72-81239	0-88318-105-3
No. 7	Exploring the History of Nuclear Physics (Brookline, 1967, 1969)	72-81883	0-88318-106-1
No. 8	Experimental Meson Spectroscopy - 1972 (Philadelphia)	72-88226	0-88318-107-X
No. 9	Cyclotrons - 1972 (Vancouver)	72-92798	0-88318-108-8
No. 10	Magnetism and Magnetic Materials - 1972 (2 parts) (Denver)	72-623469	0-88318-109-6
No. 11	Transport Phenomena - 1973 (Brown University Conference)	73-80682	0-88318-110-X
No. 12	Experiments on High Energy Particle Collisions - 1973 (Vanderbilt Conference)	73-81705	0-88318-111-8
No. 13	π-π Scattering - 1973 (Tallahassee Conference)	73-81704	0-88318-112-6
No. 14	Particles and Fields - 1973 (APS/DPF Berkeley)	73-91923	0-88318-113-4
No. 15	High Energy Collisions - 1973 (Stony Brook)	73-92324	0-88318-114-2
No. 16	Causality and Physical Theories (Wayne State University, 1973)	73-93420	0-88318-115-0
No. 17	Thermal Expansion - 1973 (Lake of the Ozarks)	73-94415	0-88318-116-9
No. 18	Magnetism and Magnetic Materials - 1973 (2 parts) (Boston)	59-2468	0-88318-117-7
No. 19	Physics and the Energy Problem - 1974 (APS Chicago)	73-94416	0-88318-118-5
No. 20	Tetrahedrally Bonded Amorphous Semiconductors (Yorktown Heights, 1974)	74-80145	0-88318-119-3
No. 21	Experimental Meson Spectroscopy - 1974 (Boston)	74-82628	0-88318-120-7
No. 22	Neutrinos - 1974 (Philadelphia)	74-82413	0-88318-121-5
No. 23	Particles and Fields - 1974 (APS/DPF Williamsburg)	74-27575	0-88318-122-3
No. 24	Magnetism and Magnetic Materials - 1974 (20th Annual Conference San Francisco)	75-2647	0-88318-123-1
No. 25	Efficient Use of Energy (The APS Studies on the Technical Aspects of the More Efficient Use of Energy)	75-18227	0-88318-124-X
No. 26	High-Energy Physics and Nuclear Structure - 1975 (Santa Fe and Los Alamos)	75-26411	0-88318-125-8
No. 27	Topics in Statistical Mechanics and Biophysics A Memorial to Julius L. Jackson (Wayne State University, 1975)	75-36309	0-88318-126-6